Engaging with Environmental Education through the Language Arts

This creative volume demonstrates the urgent importance of engaging students cognitively and affectively with the climate crisis and environmental education, underpinning the vital role the language arts play in expanding this engagement for a better future.

Moving beyond the basic modalities of English, chapters written by an internationally diverse group of contributors advocate for the integration of language arts with environmental education through broad representation of creative subdisciplines: drama, visual literacy, philosophy, poetry, student voice and more. These subdisciplines are explored to suggest the context in which environmental degradation, forest ecologies, carbon literacy and indigenous knowledges are taught, further helping students to develop a comprehensive view of how they can effect change. Ultimately, the book makes a compelling argument by emphasising the significance of interdisciplinary learning in fostering a holistic understanding of environmental issues.

This volume will appeal to scholars, researchers and postgraduate students in the field of environmental and sustainability education, English and literacy/language arts and teacher education more broadly. Undergraduate students, policymakers, environmental educators and curriculum designers may also benefit from this volume.

Nicholas McGuinn is an Honorary Fellow in the Department of Education, University of York, UK.

Amanda Naylor is Senior Lecturer in English and Education, University of York, UK.

Routledge Research in Education, Society and the Anthropocene

This series offers a global platform to engage scholars in continuous academic debate on key challenges and the latest thinking on education and society and the role of education in the Anthropocene. It provides a forum for established and emerging scholars to discuss the latest debates, issues, research and theory across the field of education research relating to society and the Anthropocene.

Rethinking Education in Light of Global Challenges
Scandinavian Perspectives on Culture, Society and the Anthropocene
Edited by Karen Bjerg Petersen, Kerstin von Brömssen, Gro Hellesdatter Jacobsen, Jesper Garsdal, Michael Paulsen and Oleg Koefoed

Ecosophy and Educational Research for the Anthropocene
Rethinking Research through Relational Psychoanalytic Approaches
Alysha J. Farrell

Justice and Equity in Climate Change Education
Exploring Social and Ethical Dimensions of Environmental Education
Edited by Elizabeth M. Walsh

Education as Human Knowledge in the Anthropocene
An Anthropological Perspective
Christoph Wulf

Engaging with Environmental Education through the Language Arts
Interdisciplinary and Creative Approaches to Fostering Ecoliteracy
Edited by Nicholas McGuinn and Amanda Naylor

Engaging with Environmental Education through the Language Arts

Interdisciplinary and Creative Approaches to Fostering Ecoliteracy

Edited by Nicholas McGuinn and Amanda Naylor

Routledge
Taylor & Francis Group
LONDON AND NEW YORK

First published 2025
by Routledge
4 Park Square, Milton Park, Abingdon, Oxon OX14 4RN

and by Routledge
605 Third Avenue, New York, NY 10158

Routledge is an imprint of the Taylor & Francis Group, an informa business

British Library Cataloguing-in-Publication Data
A catalogue record for this book is available from the British Library

ISBN: 978-1-032-61555-4 (hbk)
ISBN: 978-1-032-63451-7 (pbk)
ISBN: 978-1-032-63452-4 (ebk)

DOI: 10.4324/9781032634524

Typeset in Galliard
by Deanta Global Publishing Services, Chennai, India

Contents

List of editors and contributors

Martin Braund is an Honorary Fellow in the Department of Education at the University of York in the UK and Professor and Research Associate in the Faculty of Education at Nelson Mandela University in South Africa. Martin has over 150 publications, and his books, chapters in books and journal articles are internationally known and frequently cited. He has a lifelong interest in how the arts interact with science, developed through his work at Bretton Hall College of the University of Leeds and as Director of the Science and Drama Research Group in Cape Town. He has worked as a consultant in Brunei, Rwanda, Chile and Norway and as keynote speaker and conference presenter in over 30 countries.

Ian Davies is Professor Emeritus, University of York, UK. He is the author of many books and articles, principally about citizenship education. His experience includes being Director of the Centre for Research on Education and Social Justice at the University of York (2007–2021); Visiting Professor at The Education University of Hong Kong; Fellow of the Society of Educational Studies; Fellow of the Japan Society for the Promotion of Science; and "expert on education for democratic citizenship" for the Council of Europe.

Lynda Dunlop is Senior Lecturer in Science Education in the University of York Science Education Group (UYSEG) and Director of Education for Environmental Sustainability at York. With a background in teaching science and philosophy in schools in the UK and internationally, her research interests focus on the philosophical, political and environmental dimensions of science education, and on public engagement with new science, in particular on how to enhance democratic processes through science and education. See www.york.ac.uk/education/our-staff/academic/lynda-dunlop/.

Dr. Clare Echterling is Lecturer in English at Caldwell University whose research focuses on the environment in children's and young adult literature and teaching in the environmental humanities. She has forthcoming essays in the *Oxford Handbook of Young Adult Literature* ("Intergenerational Justice and Young Adult Literature") and the MLA's *Teaching the Literature of Climate Change* ("Decolonizing Climate Knowledge: Teaching Kathy Jetñil-Kijiner's Poetry in an Online Global Environmental Literature Course").

Miguel Farías is Full Professor of English, Applied Linguistics and Discourse Studies at Universidad de Santiago de Chile. His research agenda encompasses research articles, edited books and book chapters on literacy, multimodality, second language learning and critical discourse studies published in Latin America and Europe. He is the Editor of *Arboles y Rizomas*, an academic journal in literary and linguistic studies.

Tanya Gautam is Doctoral Researcher at the Multidisciplinary Environmental Studies in the Humanities (MESH) research hub at the University of Cologne, Germany. As part of the research project on "*Eco-Critical Literacy in Musical and Literary Practices of Cultural Education*" funded by Germany's Federal Ministry of Education, she explores informal and non-formal practices of cultural education, focusing in particular on ways of cultivating multispecies community futures through music and literature-based practices. She is the founder of *Moonlit Brooks* – a spoken word poetry project which also curates and organises multidisciplinary cultural events in Germany. In the past, she has worked as a Research Associate at the European University for Well-Being, a kindergarten educator, and a journalist.

Katharina Glas is Professor of English Language Pedagogy at Pontificia Universidad Católica de Valparaíso, Chile. Her research interests include teacher autonomy, learner motivation and the overlap between intercultural and environmental contents in ELT. She has published several articles in books and journals such as the *ELT journal*, *ecozon@*, *Innovation in Language Learning and Teaching*, and *Teaching and Teacher Education*, as well as the book *Teaching English in Chile: A Study of Teacher Perceptions of Their Professional Identity, Student Motivation and Pertinent Learning Contents* (Peter Lang, 2013).

Mark Harvey, Ngāti Toa, Ngāti Raukawa, is an artist and researcher, focussing on performance art/live art and choreography, in addition to arts with science awareness social science with science awareness, arts advocacy, mātauranga Māori, curating, collaborating and transdisciplinary. His research is centralised around connecting with communities and building public awareness around ecological issues, social justice and related general-level mātauranga Māori perspectives.

Christina Houghton (she/her) is a pākehā interdisciplinary artist, curator, researcher based in Auckland/Aotearoa New Zealand. She specialises in ecological performance, somatic dance, costumes and critical performance studies as well as writing and teaching. Her research is in community and education settings exploring dance and somatic practices in relation to social, ecological issues, health and well-being. She also co-curated the 18-month festival About Walking with Te Uru gallery.

Karen Kerr (B.Ed. Hons, Ph.D., FHEA) is Lecturer in Education in the School of Social Sciences, Education and Social Sciences at Queen's University Belfast. She works in initial teacher education and co-ordinates the science programme, specialising in Biology. Her primary research areas are education for sustainability, climate change and outdoor learning with children and young people.

Uwe Küchler has been Professor for the Teaching of English as a Foreign Language (TEFL) at Eberhard Karls Universität Tübingen, Germany, since 2016. Before his current appointments, he taught at the Universities in Bonn, Halle-Wittenberg and Dortmund. Küchler studied at Berlin's Humboldt-Universität (Germany), the University of London's Goldsmiths' College (Great Britain) and at Georgetown University in Washington DC (USA). Currently, Uwe Küchler is finalising the student handbook *Environmental Literacy and the Teaching of English* to be published with Narr publishers in 2024.

Debbie Larkins (Te Rarawa; Ngāti Kahu) is a student at Waipapa Taumata Rau, The University of Auckland, Aotearoa New Zealand where she is working towards her Master of Science degree specialising in Environmental Management. Her research examines the underlying motivations, processes, engagement and outcomes of citizen science when applied in different cultural contexts. She has been actively involved in the Tuākana Programme and holds the role of Kaiāwhina Māori within the School of Environment where she works to promote Māori and Pasifika success in tertiary science education.

Sasha Mattewman started her career as an English teacher in Bristol, UK. She went on to lead the English course for initial teacher education at the University of Bristol from 2005 to 2012. After moving to Aotearoa New Zealand, she worked as a Senior Lecturer in education at the University of Auckland and led a TLRI research project investigating literacy and environmental identities (2016–2018). Her book, *Teaching Secondary English as if the Planet Matters* (2011) was the first in the Planet Matters series for Routledge. Sasha currently teaches English at Titirangi Rudolf Steiner School in Auckland.

Billy McClune, during a varied career as a teacher, science educator, advisor and researcher, has encouraged an experience-based approach to learning. Prior to his appointment to Queen's University Belfast in 1997, he taught Physics at secondary level. During his university career he has co-ordinated PGCE science education, managed the initial teacher education programme and served as Director of Education. Science literacy has been an enduring curiosity with several school-based studies focusing on science in the news media. These studies have been informed and complemented by periods of study and school-based observation in Aotearoa New Zealand, Canada and the United States.

Jennifer MacDonald is Assistant Professor in the Faculty of Education at the University of Regina (Saskatchewan, Canada). Emerging from experiences guiding students in outdoor education courses, her research focuses on ethical relationality and how students make meaning of the living world. As a non-Indigenous educator in Canada, she is committed to modelling how all learners can dialogue with topics of truth and reconciliation to understand themselves, and to interact with others, in ways that challenge colonial logics and systemic racism.

Marie McEntee is a social scientist in the School of Environment at the University of Auckland, Aotearoa New Zealand with teaching and research interests in science/society interactions particularly relating to complex socio-environmental issues. Her research principally focuses on public engagement in science, science communication and science and technology studies (STS).

Dr Nicholas McGuinn has over 40 years' experience of working in schools and in university departments of education. His teaching, research and publication interests include English and drama pedagogy. He was recently the Ottilie Wildermuth Visiting Professor at the University of Tübingen, Germany, and is currently an Honorary Fellow in the Department of Education at the University of York. His most recent publications include a co-edited book on drama and citizenship education and a contribution to a collection of essays published to mark the centenary of the Newbolt Report (both published by Taylor Francis).

Molly Mullen is Senior Lecturer with professional experience producing theatre education, applied theatre and community arts projects in the UK and Aotearoa New Zealand. Her research examines the opportunities and constraints experienced by artists and arts organisations as they work towards social change within particular funding and policy contexts. Her book *Applied Theatre: Economies* examines the ways socially committed theatre makers fund, finance or otherwise resource their work.

Dr Amanda Naylor is Senior Lecturer in English and Education at the University of York. Her main research interests are the teaching of literature and poetry, teacher training and digital pedagogy. Amanda is currently the Programme Leader for the B.A. English in Education and Chair of the Scholarship Forum at the University of York. Amanda has experience of teaching and management in Secondary, Further and Higher Education settings. Amanda was invited by the DAAD (Deutscher Akademisher Austauschdienst) to take up the Ottilie Wildermuth Chair in the Teaching of English Language and Literature at the University of Tübingen, Germany, in 2019 and is currently involved in researching the reading experiences of undergraduate students in different national contexts.

Hannah Neville is Drama Educator and Masters Student at the Faculty of Education and Social Work, Waipapa Taumata Rau, the University of Auckland, Aotearoa New Zealand.

Hege Emma Rimmereide (her@hvl.no) is Professor of English at Western Norway University of Applied Sciences. Her research focus ranges widely, such as ecocriticism, intercultural competence, reading and writing and visual literacy, and the use of digital learning tools. Her latest publication is "Ecocritical dialogues – Enabling sustainability competency and environmental awareness within teacher education" (2023) in *Environmental Education Research* and "Developing environmental awareness in English teaching through literature" (2024) in *Current issues in English Teaching.*

Dr Jennifer Roberts is a trained teacher and works in the field of initial teacher education, specialising in English Language, English Literature and Literacy at Queen's University, Belfast. Jennifer's research interests revolve around effective education, reading and literacy. Jennifer has designed, implemented and evaluated a range of literacy programmes which seek to improve critical thinking and understanding of texts, including traditional printed material, visual forms of text and other types of media.

Jelle de Schrijver is a biologist and philosopher. He is Assistant Professor at Antwerp University (Belgium). He teaches the didactics of philosophy in the Educational master programme. In his research he focuses on the role of philosophy and philosophical dialogues in Education for Sustainable Development, science education and media literacy. https://www.uantwerpen.be/en/staff/jelle-deschrijver/

Dr Ghazal Shaikh is Assistant Professor at the University of Sindh, Pakistan. She completed her Ph.D. in education of citizenship through fiction at the University of York, UK, in 2017. Her research interests include education in literature, education and citizenship, and education in citizenship through fiction.

Ariane Craig Smith is an arts curator, researcher and mother based in Auckland, Aotearoa New Zealand. She has a Masters in Art History from the University of Auckland, and is a founding trustee and co-curator of The Kauri Project. A career working across public museums, galleries and arts festivals has fuelled a passion for creative dialogue – between artworks, between art forms, between artists and audiences, between art and other disciplines.

Dr Erin Spring is Associate Professor in the Werklund School of Education at the University of Calgary. She holds a B.A. (Honours) in English from Trent University, Ontario and a B.Ed. from Queen's University, Ontario. Erin was a classroom teacher in London, UK, before returning to graduate studies. She earned an M.Phil. and Ph.D. from the Faculty of Education at the University of Cambridge. Broadly speaking, Erin's interdisciplinary research focuses on young people's literacies, texts and cultures. Drawing on a range of methodological approaches, Erin's research seeks to understand the ways in which young people make sense of their identities through reading, writing and art.

Her research projects are united thematically by a shared investment in stories and storytelling as a way of articulating identity development, with a particular focus on the influence of place.

Peter Sutoris is Lecturer in Climate and Development at the University of Leeds. He is the author of *Visions of Development* (Oxford University Press, 2016) and *Educating for the Anthropocene* (The MIT Press, 2022). His research focuses on imagination of alternative futures, cultures of degrowth and activist pedagogies.

Introduction

Nicholas McGuinn and Amanda Naylor

In 2002, Paul J. Crutzen published a brief but influential paper called *Geology of Mankind*. In this paper, Crutzen argues that "it seems appropriate to assign the term 'Anthropocene' to the present, in many ways human-dominated, geological epoch" and notes the serious and accelerating environmental damage inflicted on the planet by "only 25% of the world population" since the start of the European Industrial Revolution (Crutzen 2002, p. 23). Crutzen is clear about who we might need to turn to in order to address this global problem. It is "scientists and engineers", he declares, who must "guide society towards environmentally sustainable management" (2002, p. 23). The contributors to this book – including those who work as science educators – feel strongly that Crutzen's chosen saviours cannot be left to provide that guidance on their own. As Uwe Küchler argues in Chapter 4:

> The enormity of climate change scenarios … not to mention the multiple environmental, socio-political and cultural disruptions that will result, cannot be approached with facts, figures or rationality alone.

Before considering what we as language arts specialists might wish to say to Crutzen's scientists and engineers, it is necessary first to explain how we intend to describe the term "language arts" for the purpose of this book.

Küchler notes the potential of the environmental humanities to "address sustainability and global challenges", and it is within the environmental humanities that we seek to position this book for several reasons. First, we are interested in Ursula Heise's account of the environmental humanities as a "new interdisciplinary matrix" designed to "combine humanistic perspectives and methods that have already developed in half a dozen or so disciplines over the last four decades" (Heise, 2017, p.1). As the subtitle of our book suggests, the adjective "interdisciplinary" is important to our contributors and we have applied it to our understanding of the language arts. In Chapter 3, Martin Braund offers an extremely comprehensive definition which includes: "creative writing, dance, drama, drawings, fabric, films, models, music, literature, natural samples (for example leaves, mud, rocks, soil and wood), painting, photographs, poetry, sculptures and videos". Examples from most of these – but by no means all – can

DOI: 10.4324/9781032634524-1

be found in this book. We appreciate that the examples included in Braund's list but excluded from ours (fabrics, natural samples and the embodied practices of dance) deserve a book to themselves – and a fascinating and important book it would be. What we have added, however, are contributions from experts in teaching English as an additional language (who incidentally describe how, in their teaching, they use many of Braund's text types as well as others he does not mention) and philosophy (Chapters 4, 11 and 6, respectively).

Where possible, we have also tried to encourage the interdisciplinary principle by including chapters co-written by colleagues working in different educational areas and/or with different areas of expertise (for example: Chapters 7, 8, 9, 10 and 13). In the same interdisciplinary spirit, the book also encourages a range of approaches to the term "Environmental Education". Küchler offers a working definition in Chapter 4:

> Environmental education is concerned with the systematic and thoughtful integration of issues related to nature, the environment or sustainability into educational endeavours. The goal of environmental education is to develop an appreciation and respect for the natural world, not only in students within the school system, but in citizens of all ages and occupations. In this way, it is hoped, societies can be steered toward a more sustainable, secure and peaceful future.

Contributors have chosen to approach the issues raised in that quotation from a number of different perspectives. In Chapter 3, Braund for example describes the history and development of Earth Education in the United States, and Arts-Based Environmental Education in Finland; while also writing in a historical vein, Billy McClune, Karen Kerr and Jennifer Roberts in Chapter 13 trace the earliest news report on global warming to a newspaper article from 1912. Moving forward a century, Katharina Glas and Miguel Farías (Chapter 11) explore the strengths and possible weaknesses of UNESCO's 2017 learning objectives for Education for Sustainable Development. Peter Sutoris (Chapter 12) takes the opportunity provided by what he prefers to call Environmental and Sustainability Education to position his work as an educator and researcher around "the idea of imagination of alternative futures". Erin Spring and Jennifer MacDonald (Chapter 7) argue the case for replacing the term "environmental education" with "place-based learning" in order to encourage a "process of *reading-with* to open possibilities for dialoguing with the local ecologies alive where we live". On a sobering note, Sasha Matthewman (Chapter 1) reminds us that diverse approaches to "environmental thinking" might not always be positive:

> Environmental thinking covers a full spectrum of possible political positions from a very conservative "business as usual" (light greening) to dark green "deep ecology" perspectives that put nature first above all human interests.

This book is *not* in favour of encouraging "business as usual". The contributors are highly aware of the existential dangers threatening the planet – Lynda Dunlop and Jelle de Schrijver's account in Chapter 6 of how "six of the nine planetary boundaries within which humanity can thrive for generations have been transgressed", or Glas and Farías' reference in Chapter 11 to "global boiling" – make that point clear. How educators working in the language arts process this information and choose to share it with the students in their care (the responsibilities implied by the word *care* are emphasised in Chapter 13 by McClune, Kerr and Roberts) is a major concern of this book and we will return to it later. For the moment, we wish to focus, in this context, on a second, fundamental, principle which we share with environmental humanists. "The environmental humanities", Heise argues, "envision ecological crises fundamentally as questions of socioeconomic inequality, cultural difference, and divergent histories, values, and ethical frameworks" (Heise, 2017, p. 2). That point is powerfully made throughout this volume, not least by Clare Echterling in Chapter 2 – a chapter strategically placed near the beginning of the book. It is revisited and expanded in Ian Davies' Conclusion, where he writes: "the issues of power and authority are centrally relevant to the environment and the language arts. Who speaks and to what effect must be recognised as being of central importance".

Questions of "who speaks?'" and "what voices are listened to?", the call for attention to Heise's "divergent histories, values, and ethical frameworks", feature heavily throughout the book, not least when Tanya Gautam makes Voice the focus of Chapter 5, where she explores ecopoetry written by young students she has worked with in order to consider how they engage with issues of growing biodiversity loss and species extinction as well as the vulnerabilities amongst human and non-human sharers of the planet. Issues of voice feature prominently in *The Routledge Companion to the Environmental Humanities* (Heise, Christensen & Niemann, 2017). We were particularly struck by statements from two of that volume's contributors, Deborah Bird Rose and Thom van Dooren:

> One of the great challenges for the environmental humanities is that of developing modes of response that hold on to a curious attentiveness to the lives and deaths of others in a world in which so much is slipping away.

And again: "An ethical response takes an interest in what matters to another rather than reading one's own positioning on to them" (Rose & van Dooren, 2017, p. 126, 125). In Chapter 5, Gautam extends this awareness of "what matters to another" by widening the definition of "voice" beyond the human to include the earth's "sonic diversity" in all its polyphonic richness.

We are therefore very pleased that our language arts contributors offer perspectives and expertise from a range of countries around the world, several of which, unfortunately, have experienced the dubious legacy either of Spanish colonialism (Chapter 11) or, as Davies puts it – quoting Sarah Bernstein – its English variant, which, she robustly declares, brought with it the "arrogance of English speakers, the way they contrived, by virtue of their tongue, to bring destruction with them, wherever they went in the world" (Bernstein, 2023, p. 64). More than this, those contributors in turn celebrate a diverse range of other voices. On many occasions, they are the voices of the young people, from primary school level to the higher education sector, who will inherit the environmental legacy we bequeath them (see for example, Chapters 1, 5, 6, 8, 9, 10 and 11). Elsewhere, they are the voices of global activists, often from within Indigenous communities. In Chapter 2, for example, Echterling reviews a series of contemporary environmental texts suitable for classroom use across the age ranges and created by writers from, amongst other countries, India, Palestine, Russia, Senegal and Uganda. In Chapter 5, Gautam engages with Chamorro poet Craig Santos Perez's exploration of species extinction in the poem *The Last Safe Habitat*. In Chapters 7 and 11, respectively, Spring and MacDonald and Glas and Farías pay tribute to the work of Potawatomi scholar and botanist Robin Wall Kimmerer. Both chapters, as Spring and MacDonald put it, "see it as valuable to bring Elders and storytellers to place alongside us to share traditional knowledge". In the case of Chapter 7, Spring and MacDonald work with Kainai scholar Tiffany Prete (Apooyak'ii) to re-interpret in a Canadian context Robert Macfarlane and Jackie Morris' 2017 text *The Lost Words: A Spell* Book (a book cited by several of our contributors). In Chapter 11, Glas and Farías draw on interviews with Elicura Chihuailaf, physician, poet and member of the Mapuche nation, to enrich their forest stewardship project. In Chapter 12, Sutoris describes the workshops he has held in India and South Africa, offering young people the opportunity to make films on subjects of their choice. In Chapter 13, McClune, Kerr and Roberts explore the ways in which young people access information about climate change through print, visual and social media. A particularly comprehensive example of inclusive approaches to voice is offered in Chapter 10, where Molly Mullen and her team of colleagues weave not only Māori Indigenous expertise and ways of knowing but the language itself into their account of their own, drama-focussed forest stewardship project (Chapters 10 and 11 provide fascinating comparisons and counterpoints).

The reasons why we began the Introduction by quoting Küchler's contention that "climate change scenarios … cannot be approached with facts, figures or rationality alone" should already be becoming clear. However, it is time now to address more specifically the question of how we might engage – the active verb is significant – with environmental education through the language arts in ways that encourage interdisciplinary and creative approaches to ecoliteracy. It is no surprise that so many (but, significantly, not all) of our contributors advocate interdisciplinary engagement. The most obvious reason is that without language we would be unable to comprehend, mediate or organise a response to the natural world and the current planetary crisis at all. Hence so many of the book's contributors take pains to articulate in detail the cognitive language skills required to engage with the issues raised in their respective chapters. It is striking, too, how many of these are replicated across different subject areas. For example, Surtoris, Glas and Farías and McClune, Kerr and Roberts all offer precise guidance on the cognitive language skills needed to deconstruct what Sutoris calls the "visual soup" of static and moving images. Dunlop and de Schrijver, in their account of the kinds of "categories talk" associated with philosophical dialogue, reference the higher-order literacy skills described by Cassidy and Christie (2013) as being required in order to participate constructively in such exchanges: "perspective taking, clarification, hypothesising, speculation, definition and differentiation of words, providing examples through the use of analogy, simile and metaphor, introducing new ideas and expressing doubt." All of the points made here – especially the references to analogy, simile and metaphor – would resonate with the kind of interactive and creative scientific pedagogical practice described in Braund's chapter. Braund, like Dunlop and de Schrijver (see their advocacy of "slow thinking" in Chapter 6), is concerned to address the cognitive challenges which can impede young people's access to a particular academic discipline – in his case, the sciences. Braund sees the development of language skills – particularly an appreciation of analogy, simile and metaphor – as a means of providing a bridge between the discourses of daily life and the more academically focussed discourses of the science classroom, where a word like "force" for example might have a very different meaning from its street definition. The drama activities Braund uses to engage with issues like this would not be out of place in Mullen and her team of colleagues' forest stewardship project described in Chapter 10. From a philosophical perspective, Dunlop and de Schrijver are also concerned, like Braund, to interrogate the connotations of words which have different everyday and academically specific interpretations. Just as Braund wishes to tease out the meanings of words like "force" or "energy", so Dunlop and de Schrijver are interested in terms like "truth", "reality" or "ecology". In Chapter 7, similarly, Spring and MacDonald interrogate the ethical and cultural implications of declaring an already inhabited landscape to be a *terra nullius* and thus ripe for colonisation.

Making sense of the "visual soup". Teaching the higher-order literacy skills. As Echterling's chapter and Davies' assertion that "issues of power and authority are centrally relevant to the environment and the language arts" suggest, the contributors to this book are committed to more than inculcating basic cognitive skills. Each chapter is concerned with helping young people develop *critical* literacy. In this context, Glas and Farías pose a fundamental question: "Up to what point do the cognitive perceptions created by our native languages permit us to overcome barriers in our thinking and embrace new ways of perceiving and interacting with the environment?" This question is explored in Chapter 5, where Gautam considers how translating her students' poems from German into English highlights ambiguities and nuances of meaning. Glas and Farías quote Robin Wall Kimmerer's observation that "It is impossible in our language [Potawatomi] to speak of other living beings as 'its'" (Tippett & Kimmerer, 2016). At this point, it might be constructive to reflect on Franz J. Broswimmer's 2002 description of *Homo esophagus colossus* – that ravenous creature who, as so many of our contributors point out without naming him directly, has for so long engaged in an exploitative and abusive relationship with other human and non-human sharers of the planet. In terms specifically relevant to this book, Broswimmer suggests that:

> Our capacity as a social species to transform nature dramatically increased during [the] early phase in human social evolution due to the development of language and the associated expansion of our symbolic and social organizational capacities.
>
> (Broswimmer, 2002, p. 23)

He continues:

> culture and language have enormously expanded the range of human possibilities. By means of our uniquely expanded biological capacity for culture, *Homo sapiens* acquired the awesome power to impose itself on nature from within. But this power is a double-edged sword: it both creates and destroys.
>
> (Broswimmer, 2002, p. 28,)

For Broswimmer, language – and literacy, with its drive to quantify and record ownership of the earth – mark the beginning of a fundamental split between human beings and the natural world. It is important, as part of this book's critical literacy project, to help young people think about this and to appreciate that the language arts can be applied to positive and negative purposes. They can be used to normalise what Matthewman described earlier as "business as usual" as well as to promulgate "'deep ecology' perspectives that put nature first above all human interests." They can be used to justify "greenwashing" (Chapter 6) or, as Echterling argues, to "place the burden on youth and their families to save the planet at home by taking shorter showers or practising Meatless Mondays, but not through civic engagement and public forms of collective activism that can inspire widespread change". In extreme cases, they can promote the sinister cause of *eco-fascism* (Chapter 2, Conclusion). As the quote from Bernstein suggests, English bears a particularly heavy burden in this respect, not least because, as Glas and Farías point out, in many parts of the world it has become the *lingua franca* through which information about environmental issues is communicated.

"Ecocritics" Matthewman writes, "think or teach outside or they at least look out of the window". In terms of this book, the statement can be interpreted literally and metaphorically. Whichever interpretation is chosen, the core point is that our contributors all place confidence in young people's capacity for agency, for taking responsibility for learning into their own hands. As Dunlop and de Schrijver put it, "the accessibility of philosophical dialogue for the consideration of controversial or contested topics" exists "even amongst very young children". Each chapter in the book, therefore, offers practical examples of young people working either literally or metaphorically "outside". In Chapter 7, for example, Spring's English Literature students accompany their reading of Susannah Moodie's (1852) *Roughing it in the Bush* with a walk "along the Otonabee River, beyond their classroom walls – the same river depicted in Moodie's text". In Chapter 9, Amanda Naylor's pre-service teachers explore the flora and fauna of the East Yorkshire coastline in order to collate material for their place poems. The "outside" might be located in a virtual space – as in Chapter 8, where Hege Emma Rimmereide and Ghazal Shaikh describe how they used the affordances of technology to enable undergraduates from England, Norway and Pakistan to explore together the environmental concerns of William Golding's 1954 novel *Lord of the Flies*. With the help of a teacher "facilitator" whose task is to challenge student thinking, Dunlop and de Schrijver's secondary students engage in face-to-face dialogue about tough philosophical and ecological questions such as "Would it be bad if people went extinct?" or "Can one live without harming the earth?"

Thinking about the "window" in more metaphorical terms, the students involved in the projects organised by Matthewman, Gautam, Glas and Farías and Mullen's team of colleagues all participate in active research projects which culminate in shared creative writing responses, presentations and, in the case of Chapter 10, wider dissemination of information and material through the community by means of technology and film. McClune, Kerr and Roberts' students engage in the science classroom with what any teacher of English would recognise as a series of *directed activities related to text* (DARTS), while Braund's students explore and share their understanding of science concepts through "poems, rap singing, poster pictures and even a short acted-out playlet". Erik De Corte (In De Corte, Verschaffel, Entwistle & van Merriënboer, 2003) argues that one of the characteristics of a powerful learning environment is that students possess the necessary metalanguage to engage at a truly purposeful level with the subject they are studying. Chapters 8 and 9 both provide examples of what happens when young people are invited to explore and discuss profound environmental issues without that metalanguage. It is only by deploying the lens of critical literacy to deconstruct otherwise superficially understood concepts such as "global warming", "ecocide", "eco-fascism" or "loss literacy" that informed engagement can take place.

All the book's contributors are committed to encouraging student agency and critical literacy through engagement with creative and interactive learning experiences. However, we noted earlier that not all of them are prepared wholeheartedly to endorse an interdisciplinary, cross-curricular pedagogy. Küchler identifies the problem:

> All of of these issues are very pressing and the urgency is undeniable. However, they are not in any obvious way consistent with the principles and objectives, content or methodology of language education. Relying

> on the facts and figures of environmental change not only reiterates the contents already dealt with in other school subjects, such as biology, geography, social studies and their bilingual branches. Arguably, the dominance of scientific knowledge in language teaching leads further away from a substantive engagement with language, culture or media communication.

Important though their role as a so-called "service subject" to environmental education might be, the language arts have far more to give than that. They can offer environmental educationalists a series of safe creative spaces – sometimes initially solitary spaces, perhaps, as in writing poetry, but always ultimately directed towards a communal endeavour and sharing – in which the imagination has the freedom to do what art is always meant to do: bring the over-familiar, forgotten or marginalised startlingly back to life by posing bold new questions and offering bracingly original ways of how, in the particular context of environmental education, we might attend to the crisis currently facing the sharers of this planet. Spring and MacDonald provide a particularly vivid example of what can be done. Describing the creative writing activities they undertook during their visits to the Bow River in 2018, they write:

> Over several seasons, our visits involved a range of literacy practices, such as letter-writing, journaling, reading alongside, and sense mapping. On our last visit, we decided to each select a text inspired by our other river visits to share with each other and the river. After reading the texts aloud, our discussion surfaced several questions: Were we involving the river in our interactions? How might the river respond or engage with us and the text? How would we ever know, since we do not speak the same language? Through pondering these questions, we realised that, in our process, we had fallen victim to objectifying place, and that we needed to consider how we were *with* place.

Küchler uses the illustrations and acrostic poems of Jackie Morris and Robert Macfarlane's *The Lost Words: A Spell Book* to encourage his students to explore language's fundamental elements of sound and rhythm while at the same time thinking about the cultural implications of losing words from the linguistic heritage. He adds an environmental dimension to the exercise by inviting the students to try to shift the focus of their anthropomorphic gaze in the way that poets such as a.rawlings or Jody Gladding attempted in works like *Wide Slumber for Lepidopterists* (rawlings, 2006) or *Translations from Bark Beetle* (Gladding, 2014):

> To further enhance the enjoyment of these poetic forms, to go beyond the visual and make greater use of other senses, the learners play with the sound structure of the environmental word and with their own voices. They read out the poems aloud and use their voices to express different moods. What does the unfolding of a fern leaf sound like? What would make their voices resemble the bickering of a magpie?

Matthewman does something similar – again through the medium of poetry – when she invites her students to write their own poems in the style of Les Murray's anthology *Translations from the Natural World* (Murray, 1992). "What," she asks, "would a shellback tick say if it could speak?" (Again, these poems, and those written by Gautam's students, make a fascinating comparison with the poems written by Naylor's pre-service teachers.)

"The question, as ever, is 'How with this rage shall beauty hold a plea?'" Quoting Shakespeare's *Sonnet 65,* the poet Seamus Heaney was thinking specifically of the violence that plagued his Irish homeland in the second half of the 20th century; but his use of the phrase "as ever" is a reminder that, though the specific question may change, the challenge it poses to the language arts remains constant. Heaney answered his own question with a quote from his fellow Irish poet, W. B. Yeats: "by offering 'befitting emblems of adversity'" (Heaney, 1980, p. 57). Writing in the context of her own subject area, Matthewman notes: "often the work in English on the environment is side on, rather than heads on. It is oblique – a subtle building of ideas and impressions that provoke or prepare for more explicit environmental thinking later, when ready". As the earlier reference to Braund's work has suggested, metaphor, simile and analogy have an important part to play here. They can help learners find a purchase on complex ideas – as when McClune, Kerr and Roberts encourage their young students to interrogate a scientific text as if it were a "News Bug":

> Each of the legs of the bug represents a critical question that can be used to interrogate the text. Pupils can assemble or disassemble the bug one question at a time. A news bug with insufficient legs is not able to stand up and might suggest that the news report lacks credibility in some key areas.

Film-making in the Pacific, Sutoris describes his attempts to find visual metaphors which might help him articulate his response to the environmental degradation he witnessed:

> Capturing the place-based loss in the Marshall Islands was what I was after. To do this, I looked for visual metaphors evocative of the sense of grief I sensed around, and within, me. It was not cerebral, fact-driven images that I hoped to produce but ones conveying a sense of atmosphere, ambiguous images to which different observers might respond differently – images that would help stimulate the radical imagination rather than suppress it.

In Chapter 5 and in the Conclusion to the book, Gautam and Davies respectively model this practice by describing how they drew upon a series of metaphors to help them conceptualise and organise their response to the issues raised.

It should be clear by now that the contributors to this book are in general agreement about what the key issues are and what specific contribution language arts specialists might make. One of the invigorating qualities of an edited collection like this, however, is that it offers opportunities to accommodate and to debate different perspectives and approaches. We want to conclude this Introduction by drawing attention to two of these points of difference. The first involves issues related to the final word of the book title: *ecoliteracy*. In Chapter 1, Matthewman describes ten characteristic features of ecocritical reading and provides examples of what these might look like in practice. Matthewman argues that "most, if not all, literary texts offer formative environmental experiences if approached through an ecocritical lens of awareness, or through the practice of ecocritical reading". Matthewman's position is powerfully endorsed by a range of ecocritical works, including Taylor (2016), Johns-Putra & Sultzbach (2022) and Head (2022). While agreeing with this general principle, Echterling, and Spring and MacDonald suggest that there is a need to rethink the literature curriculum by including new texts which offer access to fresh voices from around the globe and which have a specific focus on environmental issues – texts, moreover, which move away from fantasy depictions of the natural world, for example, or which focus on communal rather than individual endeavour. More than this, Spring and MacDonald also suggest that the new texts need a new approach to reading:

> In order to *read-with,* [sic] focus must be shifted away from text-centric pedagogy (such as book studies, character assessments, or plot devices) to attend to *who* and *where* students are in relation to the text.

The issues raised here are likely to become more pressing. In 2016, Adeline Johns-Putra published a comprehensive review of climate change in literature and literary studies whose reference section cited over 200 works of fiction, poetry, drama and ecocriticism. So much had been written, Johns-Putra noted, that it was already possible to speak of a "Canon of Climate Change Literature" (Johns-Putra, 2016, p. 272). Once a literary corpus is established to the extent that it is taught in schools and universities, all the issues associated with the selection of a literary canon come into play: whose texts are to be admitted to the curriculum? Might a decision to teach a Barbara Kingsolver novel, for example, mean there is then no space to engage with the poetry of Elicura Chihuailaf? Or vice versa? Can Mary Oliver's poem *The Kingfisher* be accepted as an example of ecopoetry, or does it fail to meet the criteria because – in the poet Juliana Spahr's resonant phrase – it got the bird "right" but failed to reference "the bulldozer off to the side that was destroying the bird's habitat" (Spahr, 2011, p. 69)? Whatever choices are made, the problem remains that, as with all canons, an atrophying orthodoxy might emerge which, rather than continuing to open up new "fields of possibility" (Appadurai, cited in Sankaran, 2022, p. 123), serves only, in Lucy Burnett's words, to reproduce "versions of the established patterns of thought" (Burnett, cited in Hamilton, 2022, p. 59).

The second point of difference we wish to consider is prompted by McClune, Kerr and Roberts' reference to educators' "duty of care" towards the young people they teach. It is probably reasonable to assume that anyone who dedicates their professional life to bringing on the next generation feels at least some degree of optimism about the future. Indeed, all the chapters in this book share a belief that something positive can be done to address the environmental challenges facing the planet. The point of difference occurs in the extent to which contributors feel their students should be exposed to the gravity of the threat facing the planet – a threat those students did nothing to create. Küchler argues that a "preoccupation with dystopian narratives and the gruelling effectiveness of despair" in the classroom leads to a sense of hopelessness and denial. It is, he continues, "of utmost importance to clarify, develop, and nurture a mindset of hope" by introducing "positive narratives and success stories, that demonstrate the power of individuals and communities to make a difference". It is significant that Küchler uses the word "enjoyment" in his description of language teaching strategies; and that commitment to positivity is seen, too, in Braund's celebration of the aesthetics of scientific exploration, Glas and Farías' pleasure in discovering "joyful alternatives that assign humans a different role in their interaction with the earth" than that afforded by Broswimmer's *Homo esophagus colossus* and Dunlop and de Schrijver's belief that engaging with complex ecological questions "can be challenging and exciting".

For McClune, Kerr and Roberts, "duty of care" means being prepared to temper that sense of optimism if their students express concerns that need to be acknowledged. Like Gautam, they emphasise "the importance of adults listening to the voice of young people and responding truthfully when providing support to help them critically decipher 'the truth' and reliable 'knowledge' about the world around them". Sometimes it is the role of the language arts to bear witness to and to mourn the ecological damage that has been done to the earth (see for example the moving conclusion to Chapter 3 or the poignant ending to the young student's poem about the extinct Tasmanian Tiger in Chapter 5: "I would have liked to save him"). In this context, Matthewman notes that "there is a spiritual dimension to reading that is linked to ritual and to reverence". By going to what she calls "the dark mountain of dystopian storytelling", we are able "to face the calamities of the present and our fears for the future together". Chapters 1 and 10 provide striking examples of this kind of "duty of care" in action.

Central to the discussion is a consideration of what language arts practitioners hope to achieve by exploring issues of environmental education with their students. Much is made throughout the book of calls to *action*; but what that word means is open to interpretation. Writing within the German context of teacher neutrality, Küchler draws upon the "Consensus of Beutelsbach" to argue that "language teaching is not aimed at taking political sides, at developing scientific, technological solutions or even political activism". Teachers have a duty to

> restrain their own political opinions and to offer alternative views, thus ensuring that the range of controversial issues is presented and that learners are equipped to analyse their own (and other) societies as well as their own contexts and positions.

Echterling however suggests that it is not enough to provide young people with the information they need in order to make informed decisions about the action they might take in the future, at a time when the state chooses to acknowledge their status as adult citizens. Drawing attention to the many youth movements currently campaigning around the world for radical action on climate change, she argues that young people are not citizens in waiting but are already capable of and entitled to active civic participation. To suggest otherwise, Echterling writes, could be regarded as patronising: "The notion that youth can or perhaps should only practise at home environmentalism stems from the model of childhood which figures them as future citizens who cannot take part in political processes". With regard to the educators' "duty of care", however, Echterling recognises that this approach raises important ethical questions. Taking action on climate issues – particularly in the current United States context – is challenging and even potentially dangerous for teachers and students alike. Davies provides helpful guidance here:

> The key question for teachers is what is being learned when students are engaged in political action, rather than what is being achieved in the form of a political objective (although, of course, there will almost always be a strong overlap between these things if we are to avoid involving young people in activities that may be politically harmful to them).

Dunlop and de Schrijver's summary of the issues is also illuminating:

> should educational endeavours not, therefore, be more focused on action? There are three responses to this. First, thinking and talking are needed in addition to action: it is not an either/or situation ... Second, philosophical dialogue can inform action, and provide a space for reflection on different types of action ... Thirdly, philosophical dialogue can play a role in normalising dialogue around climate change, developing curiosity about the current state of the planet, and exploring the potential consequences of different responses.

They continue: "By helping people study the meaning of words, express ideas in language, and reflect on ways to rethink and potentially change the world, we can open perspectives toward a different kind of teaching". That final phrase is an important one on which to conclude this Introduction. We have emphasised that the book is fundamentally optimistic in tone. Our contributors believe that the current environmental crisis offers exciting opportunities to reappraise the importance we attach to environmental education, an area which, as Davies observes, has too often in the past suffered from "the fatal combination of being low status and high risk". It is this impulse which informs Sutoris's argument that we need "to reimagine different worlds", or Mullen and her team of colleagues' call for "a curriculum that allows for inquiry-based learning and has space for Indigenous knowledge and artistic practices". Davies argues that "a conceptual framework" should lie at "the heart of an approach to the environment and the language arts". The final gift of the book is his attempt to offer a sustained account of what such a curriculum should contain in terms of content, objectives, processes, pedagogy and political action, and assessment.

References

Bernstein, S. (2023). *Study for obedience*. London: Granta.

Broswimmer, F. J. (2002). *Ecocide: A short history of the mass extinction of species*. London: Pluto Press.

Cassidy, C., & Christie, D. (2013). Philosophy with children: talking, thinking and learning together. *Early Child Development and Care, 183*(8), 1072–1083. http://dx.doi.org/10.1080/03004430.2013.773509

Crutzen, P. J. (2002). Geology of mankind. *Nature, 415*, 23.

De Corte, E., Verschaffel, L., Entwistle, N., & van Merriënboer, J. (Eds.). (2003). *Powerful learning environments: Unravelling basic components and dimensions*. Amsterdam: Pergamon.

Gladding, J. (2014). *Translations from bark beetle*. Minneapolis, MN: Wilkweed Editions.

Golding, W. (1954). *Lord of the flies*. London: Faber & Faber.

Hamilton, J. M. (2022). Weathers of body and world: Reading difference in literary atmospheres before climate change. In A. D. Johns-Putra & K. Sultzbach (Eds.), *The Cambridge companion to literature and climate* (pp. 55–68). Cambridge: Cambridge University Press.

Head, D. (2022). *Nature prose: Writing in ecological crisis*. Oxford: University Press.

Heaney, S. (1980). *Preoccupations: Selected prose 1968–1978*. London: Faber & Faber.

Heise, U. K. (2017). Introduction: Planet, species, justice – and the stories we tell about them. In U. K. Heise, J. Christensen, & M. Niemann (Eds.), *The Routledge companion to the environmental humanities* (pp. 1–10). Abingdon and New York: Routledge.

Johns-Putra, A. (2016). Climate change in literature and literary studies: From cli-fi, climate change theater and ecopoetry to ecocriticism and climate change criticism. *WIREs Climate Change, 7*(2), 266–282. https://doi.org/10.1002/wcc.385

Johns-Putra, A., & Sultzbach, K. (Eds.). (2022). *The Cambridge companion to literature and climate*. Cambridge: Cambridge University Press.

Moodie, S. (1852). *Roughing it in the bush*. London: Richard Bentley.

Murray, A. L. (1992). *Translations from the natural world*. Manchester: Carcanet Press.

Rawlings, A. (2006). *Wide slumber for lepidopterists*. Toronto, ON: Coach House Books.

Rose, D. B., & van Dooren, T. (2017). Encountering a more-than-human world: Ethos and the arts of witness. In U. K. Heise, J. Christensen, & M. Niemann (Eds.), *The Routledge companion to the environmental humanities* (pp. 120–128). Abingdon and New York: Routledge.

Sankaran, C. (2022). Climate justice and literatures of the global South. In A. D. Johns-Putra & K. Sultzbach (Eds.), *The Cambridge companion to literature and climate* (pp. 114–128). Cambridge: Cambridge University Press.

Spahr, J. (2011). *Well then there now*. Jaffrey, NH: A Black Sparrow Book.

Taylor, J. O. (2016). *The sky of our manufacture: The London fog in British fiction from Dickens to Woolf*. Charlottesville, VA: University of Virginia Press.

Tippett, C., & Kimmerer, R. W. (2016). The intelligence of plants. *On Being* [Podcast]. https://onbeing.org/programs/robin-wall-kimmerer-the-intelligence-of-plants-2022/

Part 1

Opening up the debate

1 Environmental texts in the classroom

The centrality of ecocritical reading

Sasha Matthewman

Introduction: we are only human after all

I would like to begin this sample of classroom texts and readings with an ecocritical question particularly pertinent to Aotearoa New Zealand. Can words represent a wētā? (The wētā is a benign being that looks something like a cross between a cockroach and a giant locust.) Of course, any serious attempt to understand Aotearoa New Zealand's favourite insect will necessarily be a human one, drawing either on human scientific research and/or an overactive anthropomorphic imagination. A follow-up question might be: How does a wētā see the world? There is of course no theory of mind for the unlikely looking wētā, but, despite the inability of humans to experience the world like a wētā, both scientific and literary approaches can prompt environmental understanding of this curious and ecologically significant creature. The argument for an environmental approach to English is that a literary response can speak directly to more readers beyond a specialised biological interest, shaping empathy as well as knowledge. Inevitably, we cannot understand the natural world except through our own human experiences and human representations and to anthropomorphise is one way of making the non-human world accessible and meaningful. Writers distinguish between different anthropomorphic intentions – these can be arrogant and simplistic or a genuine attempt to understand how other beings experience the world (Simons, 2002; Feddersen, 2021).

More than this there is a compelling argument that each curriculum subject should be a different lens on the environment given that we are living in an intense period of climate change and environmental crisis (see Coles, Barwell, Cotton, Winter & Brown, 2013; Matthewman, 2010; Morgan, 2012). This is not straightforward as not all subjects have a tradition of working with environmental concepts and the politics of engaging with the environment are complex.

For instance, environmental thinking does not have to be empathic and altruistic – it can be quite selfish – a desire to preserve the environments that we value for particular human activities or resources rather than a desire to preserve them for future generations or for their own sake. Environmental thinking covers a full spectrum of possible political positions and has been

DOI: 10.4324/9781032634524-3

categorised in social theory accounts from a very conservative "business as usual" (light greening) to dark green "deep ecology" perspectives that put nature first above all human interests (White, Rudy & Gareau, 2015).

From an environmental teaching perspective, this might suggest that our task as English educators is to choose texts that will engage interest in the environment, define the activities to create interest in both texts and the issues, and lastly consider how texts and mediating activities prompt evaluation of a variety of environmental positions.

What texts speak to the environment?

What texts to choose has been a fundamental debate in the field of ecocriticism,[1] a field of literary study that puts the environment first, just as feminist literary criticism takes a political orientation to the work of textual reading, putting the female perspective first. Ecocriticism takes a political and moral orientation to the world ranging over discourses of environment, nature, environmental crisis, sustainability and, more recently, the concept of the Anthropocene.[2] The principle of a shift of attention from human representation to the environment and to the non-human (or more positively the more-than-human) is the fundamental difference in the ecocritical approach to reading texts. Ecocritics have made important contributions to understanding the strategies that authors and artists use to represent the fragile and threatened "more-than-human" world. Ecocriticism is an inclusive and eclectic field ranging widely over popular cultural forms including film, virtual worlds, music, documentaries and advertising seeking "to track environmental ideas and representations wherever they appear" (Kerridge & Sammells, 1998, p. 5). The "tracking" of ecocritics is often deliberately polemical and involves identifying both helpful and damaging discourses and concepts of the environment in order to re-imagine the relationship of the human to the non-human world in more sustaining ways (Buell, 1995). This account of ecocriticism suggests connections with an extremely diverse field of studies including media, cultural studies and policy. In all of these textual activities, I suggest that it is the shift in the emphasis of the reading that is important. But while the textual study of ecocritics is very diverse, it is the ecocritical choice and treatment of specifically literary texts that I will be discussing in this chapter in relation to High School English.

One of the early "first wave" ecocritics, Lawrence Buell, argued strongly and moralistically that the texts that the ecocritic should choose ought to be drawn from nature writing and that a new canon of nature writing should be discussed and taught in classrooms. This was in part an attempt to put nature writing on the literary map so that it was valued as "literary" as well as a genuine recognition of the ecological potential of these texts. Buell and others worked to define a canon of texts that could have a positive impact on how readers engage with environmental issues, even providing a handy checklist for how to evaluate the environmental worthiness of texts (see below):

1. The nonhuman environment is present not merely as a framing device but as a presence that begins to suggest that human history is implicated in natural history.
2. The human interest is not understood to be the only legitimate interest.
3. Human accountability to the environment is part of the text's ethical orientation.
4. Some sense of the environment as a process rather than as a constant or given is at least implicit in the text.

(Buell, 1995, pp. 7–8)

Buell's testing formula positions ecocriticism as taking account of the interest of nature on its own terms and for its own sake (and this is consistent with a deep ecology approach). The canon of American nature writers scored very favourably in relation to this test of ecocritical worth and there followed a swathe of American ecocritical attention to writers such as Henry Thoreau, Mary Austin, John Muir, Leo Aldpold, Edward Abbey, Annie Dillard, Terry Tempest Williams and Barry Lopez. Problematically for many literary theorists brought up on Barthes' (1972) "death of the author", ecocritical meanings are seen as inherent in the text rather than located in the process of reading. But the success of Buell (often quoted as the father of ecocriticism) was to redraw the canon to give nature writing some status in the academy (and therefore wider reach and power to ecocritical practice in tertiary courses). This has had some success with a range of English courses now featuring ecocriticism and nature writing and the classics of environmental literature have been extensively discussed in ecocritical circles. The power of an ecocritical text is often centred on the example of Rachel Carson's *Silent Spring* (1962), a mixture of scientific fact and fable which led to the banning of certain pesticides in the United States a decade after her death in 1964.

Relatively recently there has been a resurgence in the popularity of nature writing in Britain with the rise to prominence of Kathleen Jamie, Richard Mabey, Robert Macfarlane and Mark Cocker with recent highly acclaimed work by Richard Kerridge – *Cold Blood: Adventures with Reptiles and Amphibians* (2014) – and Helen MacDonald *H Is for Hawk* (2014). This has been designated as "new" nature writing. Joe Cowley's reflection in *Granta Magazine* on the qualities of *The New Nature Writing* stresses the confrontational qualities of the writing and the awareness of threat in the context of ecological crisis (Cowley, 2008). This is writing which responds to the steady stream of concern about a multi-dimensional ecological crisis played out on local and global scales. Not only do these writers show awareness of ecological crisis but they provide a powerful corrective to the pristine wilderness trope so prevalent in American "classic" nature writing. Nature is shown to be compromised and often malevolent, but ultimately unavoidable and necessary. The accommodation is that of a "magical mundanity" in the broken nature that we co-inhabit with the more-than-human world. This is the British version of the nature epiphany – the "ditch vision" which sees a value in the local, less-than-pristine environments of "ordinary" rural and urban nature (Garrard, 2007, p. 63). Clearly these texts speak very directly and polemically to contemporary environmental questions.

But what kind of readers are paying attention to the messages from nature writing whether classic or "new"? Generally, the ones whose interest in the natural world is already highly engaged and for whom the reading of nature writing works as a kind of soothing affirmation of existing and righteous environmental values. In a way they already "speak for themselves" – they embody the ecocritical message in a fairly obvious way if the reader is prepared to pay attention to it.

This straightforward synchronicity of text and reader is unlikely to be found within the high school classroom, where students may not be particularly interested either in the environment or in reading. Moreover, the reflective, digressive and descriptive mode of nature writing is unlikely to appeal easily to a student audience brought up on fast-moving visual spectacle and digital action. Getting students to engage with literary texts at all is now a fundamental issue for the English teacher in the secondary school classroom. After all, texts cannot "speak" to the environment if no one is listening. And English teachers cannot always choose their text. The work of an English teacher is often about finding ways into the engagement with the text that they have been given rather than a text that they have chosen (such as a text mandated by the exam syllabus, or available on the departmental shelf). If the teacher is lucky enough to have a free choice of text, then this is guided principally by what is most likely to appeal to a particular group of students rather than a choice of text for its environmental message. However, it is not unusual for versions of English teaching to have an implicit social conscience and for teachers to explore the moral dilemmas represented within texts. These moral dilemmas will naturally include environmental issues – but these are not foregrounded in teaching to the same extent as issues pertaining to society and culture and the "go to" issues are still those of race and gender for English (Glotfelty & Fromm, 1996). The first step is to recognise that most, if not all, literary texts offer formative environmental experiences if approached through an ecocritical lens of awareness, or through the practice of ecocritical reading. It is the practice of ecocritical reading and teaching that is central to realising the potential of English to contribute to environmental consciousness.

Working with ecocritical principles

So how do you read a text ecocritically? What does an ecocritic do? I have found it helpful to think about this question in relation to principles of practice drawn from my reading of ecocritical books and essays. These principles first appeared in *Teaching Secondary English as If the Planet Matters* (Matthewman, 2010) and they were revised in relation to research in Aotearoa New Zealand and new developments in the field of ecocriticism (Matthewman, Rewi &

Britton, 2017). For me these principles have stood in relation to the major transition from one nation and culture to another in the move from the UK context to Aotearoa New Zealand. However, this transition has brought shifts of emphasis, in particular a stronger attention to the interdependence of culture, identity and environment. The principles are an abstraction from tertiary ecocritical practice in relation to what might then "translate" to school classroom practices of ecocritical reading. For ease of reference, the principles are reprinted in brief below:

1. Ecocriticism makes the environment and the non-human central to the interpretation of the text.
2. Ecocriticism has a related interest in animal welfare and the representation of animals.
3. Ecocriticism takes account of the environmental conditions both at the time of writing, within the setting and at the time of reading.
4. Ecocritics take a moral and/or political stance on environmental issues and may urge environmental action.
5. Ecocriticism breaks through traditional boundaries between texts and disciplines connecting readily with science.
6. Ecocritics challenge the idea of nature as only existing in language. This means a challenge to poststructuralist approaches to reading where reality (and nature) is claimed to be only constructed through language. Ecocritics tend to argue that there is a real material planet that really needs saving!
7. Ecocriticism has historically taken an interest in nature writing.
8. Ecocriticism can celebrate, or show the extent of the environmental crisis.
9. Ecocritics think or teach outside or they at least look out of the window.
10. Ecocritics think about the future in the context of the Anthropocene and our collective responsibility to the natural world.

(Matthewman, 2010, 2017).

In the following section, I will show how texts can work with these principles of ecocritical practice. Reading is always located in time and place, and unusually for a solitary activity, reading in a classroom can be a uniquely shared process of mediation, interpretation and engagement with a text. Activities in relation to a text relate to reading as the work of interpretation and exploration rather than simply decoding words on the page.

The obvious way is to approach the environmental issue head-on with texts that are self-consciously environmental in a Buell-inspired way. The "speak for" attitude of environmental texts appeals to a young person's sense of justice and fairness. They are texts on a mission to change us, or our perspective. Although often unappealingly preachy, this mode can be particularly enjoyable for teenagers with their strong polarities of justice and injustice. Anthologies such as *Earth Shattering: Ecopoems* (Astley, 2007) or *No Other Place to Stand* (Ranapiri & Hawkes, 2022) can be a helpful source of poems that shout loudly about environmental issues and can work well with spoken word angry poems or other polemical pieces.

But this polemical mode does often lead to students' writing that merely proselytises about issues of recycling and pollution or saving the whale. A more complex response is needed.

Ecocritical poetics

A more comprehensive approach to engaging with the radical otherness of nature is to deliberately expand the range of environmental knowledge in the classroom as a collective endeavour. A fairly wide cross-section of the fauna and or flora of a particular region or country can be covered within a classroom of responses. Each student can be asked to research and write about a particular creature. In one high school classroom (ages 13–14), students wrote about over 20 Aotearoa New Zealand native creatures including a weka, a fur seal, a wētā, a blue-eyed penguin, a kea, a tui, a tuatara, a kōkāpō. In another class, a range of native Aotearoa New Zealand trees and shrubs were given as subjects for writing. The poems and prose pieces came together in an anthology illustrated with careful botanical drawings and detailed paintings of the birds and creatures.

Students were asked to consider the voices and approach of three main poets: the affectionate relationship of Māori poet Hone Tuwhare to the natural places of Aotearoa New Zealand, the brutal wonder of Ted Hughes' response to British nature and Australian poet Les Murray's unique "voicing" of animals found in Australia in his anthology *Translations from the Natural World* (1992). Murray's "translations" had the most discernible influence on the students' writing, and the influence of his poems was most frequently mentioned in student commentaries. Students were asked to write from the creature's perspective and to attempt to show how it thinks, feels and acts on the world through different senses – such as the way a fish experiences the world through the porous contact between skin and water checking through their skin the chemical composition. They were asked to consider how language in Les Murray's poems reflects that imagined worldview. For example, in *Eagle Pair* (p. 15) the language is blunt and telegraphic as when the eagles "shell down", and in the poem *Spermaceti* (p. 44) the representation of the whale emphasises the use of sound to orient and navigate, while the unity of a shoal of fish in *Shoal* (p. 22) is expressed in the obsessive wordplay on eyes as both collective identity and visual perception, while the blood obsession in *Shellback Tick* (p. 40) is shown in the tactile description of feeding on blood as a kind of viscous breathing. We noted how nouns might become verbs, or marked evocative neologisms. Students were asked to notice how Les Murray uses a range of archaic or obsolete language such as "thereof" to convey how the tick is an ancient being (with all the associated ecological implications of that longevity).

In these poems, creatures speak in their own unique thought language and the poet seeks to imaginatively engage with the questions and concerns of radically other frames of mind. What would a shellback tick say if it could speak? The language presses against the reader's comprehension as the students work at "translation" and grasp at the unintelligible. The poems are extremely difficult to "translate". However, students are not asked to understand or to render the meaning back in simple and straightforward terms but to identify lines and phrases that capture an "alien" consciousness and to notice the way that rhythm and onomatopoeic soundscapes mimic and suggest elusive animal communication that strains precise interpretation. The focus is on noticing writing for the purpose of writing inspiration and poetic method inviting a change of consciousness, a learning engagement and impression that will linger in the soul memory rather than explicit commentary and analysis. Encouraging an openness to lack of comprehension and to partial insights is freeing; this is initially an appeal to feeling rather than intellect, an important principle within Rudolf Steiner Waldorf education which works with pictures and feeling before explicitly engaging the intellect.

The student's assigned creature or plant will always mean something to the writer – even if they were initially uninterested. Aotearoa New Zealand's wildlife does not include large and charismatic creatures. In reply to an invitation to name the range of Aotearoa New Zealand wildlife, one student said sarcastically "well we have a lot of birds that can't fly". One ecocritical aim of the anthology was to engage students with animals that are not often anthologised or celebrated in literary examples (as in Les Murray's attention to insects such as the shellback tick, cuttlefish and mollusc). This is also important in environmental work as it is the charismatic creatures who tend to attract funding (and saving) whereas ecological health depends on the whole ecosystem. One student chose to write about the wētā taking her inspiration from Murray's poem about the shell back tick. She puts the point simply: "I try to make the reader empathise with the Wētā by writing in first person from the Wētā's perspective".

> I'm the king of the insects as large as a mouse
> Aotearoa New Zealand forever my home
> I'm scaly and long horned
> With a terrible glare
> Hearing through my knees and breathing through my skull
> I am the regal God of ugliness

Her poem weaves in her biological and cultural research about the wētā. Her striking metaphor of "the regal God of ugliness" comes from the fact that Māori considered all wētā repulsive even naming the giant wētā after Punga the God of ugliness. This student celebrates the ugliness of the wētā as an act of defiant superiority.

Another student introduces his poem with clarity and a wry recognition of the inherent impossibility of the task of understanding animal consciousness.

> This is a poem about New Zealand's only native owl, the Morepork. The poem focuses on the owl at night as this is when they are most active, being nocturnal and hunting in the dark, and is written from the perspective of the owl. It is in fact written in the head of the owl as if the author *were* the owl. This is all guesswork, however, as I don't know what owls think about.

It is easy to feel the influence of this author's reading of Murray in his poem with its insistent repetitions, telegraphic speech and naming by metaphors of imagined animal obsessions (such as the reference to the "red life draining").

Morepork

by Cameron Le Pelley
The night is mine
I am the hunter of dark
Swoop silently listening, listening…
Food squeaking-
Food speaking
Follow food tracking, tracking…
I fold my wings and dive,
Pinpoint as I go
The rush through feathers faster, faster…
Grab the food, the life-
The red life draining
And I am living, living…
The night is mine
I am the hunter of dark
Swoop silently listening, listening…

Each student's strong individual engagement with one creature and/or one plant is important in itself and requires reading and research of non-fiction and biological texts as well as reading of poetic models (remembering that ecocritics connect readily with science). The value of the classroom is that this interest can be broadened as the poems are shared aloud and re-read as a full collected class anthology. The shared culture of a classroom and the collective oral experience that is cultivated together can become a process of enviro-cultural learning.[3] The learning builds knowledge of how environment and culture are connected (many of the creatures are named twice, once in English but originally in the Māori language with associated myths and associations. For example, the friendly fantail or piwakawaka bird is a feared harbinger of death when it flies indoors). Reading may also push towards ecocritical understanding of the broader biodiversity issues and connections between species, environments and cultural practices, if the teacher can draw this out as part of plenary discussions. The approach can encourage an ecocentric perspective that views nature as independently and intrinsically valuable, rather than merely valuable as a resource for humans. This is fundamental to all ecocritical reading (although not necessarily part of all environmental thinking). In terms of action, ecocritical reading can involve strong communication of the biodiversity value of Aotearoa New Zealand wildlife including awareness of the species which are under threat and what is being done to counter this. In this reading and response, it is possible to trace principles such as the engagement with nature writing, connections with science and a strong engagement with the problem of how to represent animals and the more-than-human moving towards a deliberate attempt to decentre the human interest as the only legitimate perspective on the world.

Responses to nature are shaped by cultural perspectives and the environmental identities that are available within that culture. In particular, the Aotearoa New Zealand obsession and anxiety about belonging, culture and identity has highlighted for me the importance of environmental identities and enviro-cultural literacy (Matthewman, 2017).

I would like to illustrate this point in relation to a popular author in Aotearoa New Zealand. Patricia Grace is an Aotearoa New Zealand Māori author who explicitly seeks to raise awareness of issues of race and culture in her writing. However, her stories show how deeply narratives of nature are wrapped up in representations of Māori as Tangata Whenua (People of the land) although environmentalism would not be the explicit focus of the story or the usual focus of the reading. The short story *A Way of Talking* (1975) is usually read within English classrooms as being "about" culture and racism but an ecocritical reading also raises the ongoing cultural conflict over attitudes to, and management of, nature (a major contemporary issue in Aotearoa New Zealand with non-native mature trees felled to plant native saplings as part of a reclaiming of an authentic enviro-cultural heritage). This particular story is narrated by Hera, who is from a rural Māori background. Her sister Rohe has recently arrived home from Auckland University and is referred to as "Rose" in the narration and "Rohe" in the dialogue to reflect her dual identity as both "Pakehafied" ("becoming" European or white) and of Māori heritage. The pivotal moment in the story occurs when Rose/Rohe meets Jane, a Pākehā[4] neighbour who has moved into the community while Rose/Rohe has been away at university. During the course of a friendly social exchange between the three women, Jane makes a casual remark which underlines their social and cultural distance as she points out her husband Alan returning and says that he has been "getting the Māoris for scrub cutting" (1975, p. 8). Rohe has a visceral reaction to the dismissive "way of talking" that Jane uses to describe her people. But reading ecocritically, Jane is also dismissive and ignorant about the land. Jane's comment about cutting scrub refers to backbreaking work to clear ground for a landowner and the scrub, like the people who cut it are, in her view, not worthy of naming. However, the meaning of "scrub" does not just depend on the type of vegetation but also depends on the environmental identity of the speaker. For instance, "scrub" might mean non-native weeds invading pasture, or firewood, shrubland, a forest in the making, native kanuka and manuka "a treasure trove of indigenous biodiversity" and a significant habitat for invertebrate fauna and forest birds including threatened species of whitehead, popokatea and fernbird, a source of honey and medicinal oil, effective erosion control and culturally valued material for tools (Ward, 2005 n.p.). If the story invites interrogation of what it means to be designated as Māori or Pākehā, it also invites questioning about how our "way of talking" implies a relationship with, and attitude to, the natural world and a sense of belonging to, and understanding of, the native fauna and flora of the land.

Similarly, the novel *The Call of the Wild* (London, 1903) raises issues of environmental understanding of place as part of enviro-cultural knowledge and practice. It is a book that speaks strongly to the environment in that students are immersed in the imagination of a frozen world and an alien environment over the course of the shared reading in class. London is extremely interested in how the harsh environment of the Yukon Valley in Canada demands respect and understanding. He shows how humans adapt to this hostile climate and how other animals are more easily and instinctively attuned to this world. He also shows how humans extract resources from the land, damage it and endanger themselves and others so that humans as a destructive force impacting on the environment is deeply felt. The characters of Hal and Mercedes and their ignorant dismissal of the conditions and the animals represent a violent imposition on the land. This connects very smoothly to the settling of a place and how newcomers lack understanding of how to live in an unfamiliar environment. The portrayal of the gold rush connects with mining in other places and the damage done to ecosystems. The arrogance of the newcomers is especially resonant in colonial contexts such as Aotearoa New Zealand. The allegoric message shows the process of adapting to become "one" with the environment as Buck, a domesticated dog, gradually adapts, recovers his ancestral instincts and blends in with his environment to answer the call of the wild. This reversion to the primitive state is not an option for humanity (unless by some hideous apocalyptic mistake). Returning to the fire, to hunter-gathering and to Buck's strange vision of original "natural" man is a mythic fantasy only available to a few (see Diane Cook's excellent novel *The New Wilderness* (2020) for that particular dystopian scenario). Humans evolve out of nature, into civilisation and according to Rudolf Steiner into further development of collective consciousness (see Gidley, 2007 for a summarised account). The question for environmentalists (and for us all) is how civilisation and consciousness can live more lightly on the earth and evolve to a higher level of stewardship.

Often the work in English on the environment is side-on, rather than head-on. It is oblique – a subtle building of ideas and impressions that provoke or prepare for more explicit environmental thinking later, when ready. Ecocritics read the environment as central to interpretation and connect with animal representation as part of this shift of attention to the more-than-human. However, the teaching of English does not mean that the environment should be constantly foregrounded but neither should it be forgotten. In this case, to write an essay on the impact of humans on the Yukon Valley ecosystem would mean that it was no longer English but a social science essay. However, the alert ecocritical English teacher can seed and suggest the environmental issues that may be picked up more explicitly in the wider curriculum within geography, history or social sciences. The emphasis in English is on the feeling response and developing the consciousness of the environment as a force and process.

Historical approaches to literary criticism tend to emphasise the social and cultural contexts of a text but seldom the particulars of place or environmental issues and attitudes. Ecocritical readings make these factors central. This may involve three layers: the environmental context that underpinned the text at the time of writing, the environmental context at the time of the setting of the text (and the representation) and the environmental context that is influencing the reading in the current moment. For example, in January 2023, Auckland in Aotearoa New Zealand was hit first by intense rainfall causing widespread flooding and landslides and then a week later by Cyclone Gabrielle which caused even more damage to the coastline as well as inland, with landslides affecting coastal and rural properties and hundreds of houses being "red stickered" as unsafe to enter. The effects of the cyclone were particularly devastating in Hawkes Bay and the Esk Valley. The start of the school year in Auckland was delayed by a week due to the unprecedented intensity of the storms and flooding. When we returned to school it was still raining and I chose to begin the Steiner certificate course with Class 11 (16–17 years old) by reading Ray Bradbury's science fiction short story *All Summer in a Day* (1954) which, with powerful poetic brevity, imagines a planet of continual rain and a "concussion of storms", where the sun shines once every seven years and only then for one hour (Bradbury, 1954, p. 122).

But the relentless rain described in the story was also happening outside the window. As I read with students, we connected with the story on a feeling and visceral level as our small wooden classroom was pounded by the rain as if it was inside an Aotearoa New Zealand waterfall in full spate. Reading the story felt like a way of processing the cataclysmic climate events that had just happened in our community and in other parts of Aotearoa New Zealand and prompted stories about experiences of the storm. Many of the students chose to write about the theme of climate change in the story. This was less a reading of the context of the time and setting and more a response to their contemporary interpretation and fears. When this story was written in the 1950s it was the prospect of space travel and alien possibilities that drove the author's imagination rather than climate change, but it seemed legitimate to find this prescience in the text. In Māori culture the past is never "finished" but shifts in relation to the present so that past, present and future are in constant negotiation and this can change our attitudes to texts as changing voices in changing contexts (Matthewman, Mullen & Patuwai, 2015). Unexpectedly texts may speak to us about the environmental context of the present out of the concerns of the past.

Certainly, I was thinking of past, present and future contexts in the writing of the school play *Creation and Command* (a devised dramatic retelling of biblical stories with eurythmy, mime and song). Descriptions of Noah's flood were inspired by Bradbury's short story to speak directly to the current feelings of the school community, many of whom suffered losses in the storms. This scene became a way of mourning and acknowledging the devastation wreaked by Cyclone Gabrielle as well as a representation of the biblical Flood. Many of the audience members commented that the references had spoken to them. There is a comfort in feeling that this hardship is something that we face together – again this is a shared moment of reading through rehearsal and performance.

> Noah: No time to waste.
> You better get on this boat if it's the last thing you do.
> The waters are rising. Fast.
> *Physical boat created with mime. All looking out to sea.*
> It rained. It rained for forty days
> And forty nights;
> With lightening
> With thunder
> With hurricane winds.
> A thousand forests were crushed under
> the rain and all the cities drowned at once
> We recited the names of the lost cities
> One by one, in sad remembrance.
> Most days we could not even see
> Our hands in front of our faces
> Just the deep and endless grey
> Water and sea stretched endlessly
> No land to be seen for mile on mile
> night and day disappeared
> We almost drowned
> In darkness, sound and water
> Noah's Daughter: Now is that a chink of light?
> The rain is easing
> The world quivers until held still.
> Noah's daughter: The rain has stopped.
> Noah: send a raven and a dove.

This collective remembering connects with a "dark mountain" response to the environmental crisis (Kingsnorth & Hine, 2009; Kingsnorth, 2017). At its bleakest, "dark mountain" is a despairing response but it can also fulfil a deep need for a ritual mourning and remembrance of the losses of the environmental crisis. Kingsnorth and Hine's "dark mountain" is a call to gather, to remember, to elegise and to gain support from each other in troubled times. In going to the dark mountain of dystopian storytelling, we face the calamities of the present and our fears for the future together. This mutual support holds some space open for light, hope and redemption. This hope may not be in the text but should be part of an ethical teaching framework. In particular, a Waldorf Steiner approach calls for a spiritual response to material concerns and the recital of verses is part of an incantatory shared reading practice that draws on the ritual and remembrance of nature as blessing. For example, the verse that begins the day in most Waldorf Steiner schools evokes cosmos and spirit grounded in the Earth. It is deliberately not given to children as a written text but is learned through collective recital (and there are many variations – even sometimes from class to class). These are the opening lines in our High School version:

> I look into the world
> Wherein there shines the sun
> Wherein there gleam the stars
> Wherein there lie the stones
> The plants they live and grow
> The beasts they feel and live
> And I to spirit give
> A dwelling in my soul

This is a hopeful affirmation of the world. The words of the verse evoke bright mountain (rather than dark mountain) recitation as a daily practice of ritual awareness of our spiritual connection to a resonant living and fragile world.

Conclusion

The response to the environmental crisis and climate change has proved that it is possible to engage with environmental texts and ideas and at the same time maintain a stubborn belief in "business as usual". The idea that small actions such as household recycling or using a tote bag will be enough to mitigate ongoing climate disasters and the looming climate meltdown is fanciful. An approach to teaching that emphasises a "stupid optimism" (Facer & Selwyn, 2021) or a "bright green approach" (White, Rudy & Gareau, 2015, p. 178) can have the effect of raising entitled eco-consumers who ostentatiously drive new expensive electric cars (with their own disastrous environmental effects) while jetting all over the world, or aspiring to. At the same time, approaches that are doom-laden can also paralyse environmental engagement and response, as in a relentlessly "dark mountain" approach to the crisis. Informed awareness of issues and actions is important but in order to actually make any difference a change of consciousness is required to make possible challenging and radical changes in lifestyle and social life. "Reconciliation ecology" has been used to describe a new awareness that everything has shifted in relation to the climate crisis (White, Rudy & Gareau, 2015). Reconciliation ecology means to mitigate, adapt and reconcile ourselves to an irredeemably compromised climate and nature.

The examples in this chapter suggest that acts of ecocritical reading might lead to changes of consciousness and shifts of perspective. This is the necessary first step toward teaching an ecocritical English capable of interrogating environmental values, identities and political positions. To return to Buell's test for choice of texts, this can be reframed as a kind of ecocritical reading test:

1. Thereader is able to show how setting is both natural and constructed, uncovering the human impact on the environment.
2. Thereader focuses on the author's representation of the more-than-human environment as a living and active presence including attention to the representation of animal "actors".
3. Thereader considers what lessons or wisdom about how humans should live on the Earth are revealed through the action and representation in the text.
4. The reader demonstrates an understanding that the environment is part of a historical process, and integrates discussion of the environmental and cultural conditions at the time of writing and the enviro-cultural setting as well as how present awareness of the environmental crisis impacts our response to the text.

A test for texts becomes a test for readers as the central contribution of English to a wider understanding of the environmental crisis. As teachers in schools, we need to find a way of reading and a way of talking about the environment whatever texts we choose to teach. But it is also worth remembering that there is a spiritual dimension to reading that is linked to ritual and to reverence. This is the undefinable source of spiritual sustenance that is to be found in many literary texts. Presence of nature, and presence in nature, is felt in the poems of Les Murray and connects with the aphoristic principle "spirit is never without matter/matter is never without spirit" included in a verse regularly recited in Waldorf schools (Steiner, 1919).

In the Aotearoa New Zealand Māori world view there is a taken-for-granted belief in "mauri", and perhaps Wordsworth's feeling for "the spirit that runs through all things" helps gesture towards the meaning of this concept (Wordsworth, 1984). Attention to mauri and to wairua (spiritual health) is imbued within many social and cultural settings in Aotearoa New Zealand including education through common practices of karakia (Māori verse or prayer). The act of reading and recitation can often be an act of prayer and reverence for the natural world. In some texts, this is felt very strongly and is indeed part of the writer's intention. Writers are attuned to the spiritual hum of language and the idea of some words and texts as sacred. Poetry is the heightened language of biblical, religious and prayer texts and recitation is part of an act of collective worship and/or meditation. It is my belief that the connection that students felt to Murray's *Translations* was a response to the spiritual and material *presencing* that he achieves in his work. The poems are an act of presencing, of making material and vivid the creatures of the world. He dedicated this collection "to the glory of God", and I read this as a belief and evocation of the presence of the spirit in all things. English teaching, when imaginative and inspirational (or perhaps simply sincere and deeply felt), can work at the soul level, helping students to see the natural material world as full of meaning, purpose and spirit. This, surely, is an ecocritical aspiration for reading both the word and the world.

Notes

1 There are dedicated introductions to ecocriticism by Garrard (2004) and Clark (2011, 2019) and chapters in most major introductions to literary theory such as the fourth edition of *Beginning Theory* by Peter Barry (2017), the second edition of *Introducing Criticism in the 21st Century* by Julian Wolfreys (2015) and the fourth edition of Andrew Bennett and Nicholas Royle's excellent *Introduction to Literature Criticism and Theory* (2009) which includes new chapters titled "eco" and "animals".

2 The Anthropocene is the concept that we have left the Holocene period of climate stability and have entered a geological era which is defined (but not controlled) by human activity.

3 This term is taken from the model of 3D Ecoliteracy developed in the three-year TLRI (Teaching Learning and Research Initiative) research project that investigated ecoliteracy across English, Social Sciences and The Arts. The project adapted the 3D Literacy model of Bill Green and Catherine Beavis (2012) to investigate eco-operational literacy, enviro-cultural literacy and ecocritical literacy across these three subject areas (Matthewman, 2016; Matthewman, 2020).

4 Pākehā is usually taken to mean white European although it is also defined as referring to anyone who is not Māori.

References

Astley, N. (Ed.). (2007). *Earth shattering: Ecopoems*. Tarset: Bloodaxe.

Barry, P. (2017). *Beginning theory: An introduction to literary and cultural theory*. Manchester: Manchester University Press.

Barthes, R. (1972). *Mythologies*. New York, NY: Hill and Wang.

Bennett, A., & Royle, N. (2009). *An introduction to literature, criticism and theory*. Harlow: Pearson Longman.

Bradbury, R. (1954). All summer in a day. *The Magazine of Fantasy and Science Fiction*, March, 122–127.

Buell, L. (1995). *The environmental imagination: Thoreau, nature writing, and the formation of American culture*. Cambridge, MA: Harvard University Press.

Carson, R. (1962). *Silent spring*. Boston, MA: Houghton Mifflin.

Clark, T. (2011). *The Cambridge introduction to literature and the environment*. Cambridge: Cambridge University Press.

Clark, T. (2019). *The value of ecocriticism*. Cambridge: Cambridge University Press.

Coles, A., Barwell, R., Cotton, T., Winter, J., & Brown, L. (2013). *Teaching secondary mathematics as if the planet matters*. Abingdon: Routledge.

Cook, D. (2020). *The new wilderness*. London: Oneworld.

Cowley, J. (2008). Editor's letter. In *The new nature writing* (pp. 7–12). London: Granta.

Facer, K., & Selwyn, N. (2021). *Digital technology and the futures of education – towards 'non-stupid' optimism*. UNESCO. https://unesdoc.unesco.org/ark:/48223/pf0000377071

Feddersen, M. M. (2021). The zoopoetics of Les Murray: Animal poetry, attentiveness and the more-than-human world. *Leviathan: Interdisciplinary Journal in English*, (7), 90–111. https://doi.org/10.7146/lev.v0i7.125212

Garrard, G. (2004). *Ecocriticism*. London: Routledge.

Garrard, G. (2007). Ecocriticism and education for sustainability. *Pedagogy: Critical Approaches to Teaching Literature, Language, Composition, and Culture, 7*(3), 359–383. https://doi.org/10.1215/15314200-2007-005

Gidley, J. (2007). Educational imperatives of the evolution of consciousness: The integral visions of Rudolf Steiner and Ken Wilber. *International Journal of Children's Spirituality, 12*(2), 117–135. https://doi.org/10.1080/13644360701467428

Glotfelty, C., & Fromm, H. (Eds.). (1996). *The ecocriticism reader: Landmarks in literary ecology*. Athens, GA: The University of Georgia Press.

Grace, P. (1975). *Waiariki, and other stories*. Auckland: Longman Paul.

Green, B., & Beavis, C. A. (2012). *Literacy in 3D: An integrated perspective in theory and practice*. Victoria: ACER Press.

Kerridge, R. (2014). *Cold blood: Adventures with reptiles and amphibians*. London: Virago.

Kerridge, R., & Sammells, N. (Eds.). (1998). *Writing the environment: Ecocriticism and literature*. London: Zed Books.

Kingsnorth, P. (2017). *Confessions of a recovering environmentalist*. London: Faber & Faber.

Kingsnorth, P., & Hine, D. (2009). *Uncivilization: The dark mountain manifesto*. Retrieved March 18, 2024, from https://dark-mountain.net/about/manifesto/

London, J. (1903). *The call of the wild*. London: Macmillan.

MacDonald, H. (2014). *H is for hawk*. London: Jonathan Cape.

Matthewman, S. (2010). *Teaching secondary English as if the planet matters*. Abingdon: Routledge.

Matthewman, S. (2016). *Tuhia ki te Ao - Write to the natural world*. Wellington: Teaching Learning and Research Initiative. Retrieved March 25, 2024, from http://www.tlri.org.nz/tlri-research/research-progress/school-sector/tuhia-ki-te-ao-write-natural-world

Matthewman, S. (2017). From place to planet: The role of the language arts in reading environmental identities from the UK to New Zealand. In D. Stevens & K. Lockney (Eds.), *Students, places and identities in English and the arts: Creative spaces in education* (pp. 1–13). New York, NY; London: Routledge.

Matthewman, S. (2020). Nation and nature in L1 education: Changing the mission of subject English. In B. Green & P-O. Erixon (Eds.), *Rethinking L1 education in a global era. Educational Linguistics*, 48. Cham: Springer. https://doi.org/10.1007/978-3-030-55997-7_11

Matthewman, S., Mullen, M. S. C., & Patuwai, T. (2015). The river talks: An ecocritical 'kōrero' about ecological performance, community activism and 'slow violence'. *Research in Drama Education, 20*(4), 442–463. http://dx.doi.org/10.1080/13569783.2015.1065726

Matthewman, S., Rewi, N., & Britton, R. (2017). Locating eco-critical literacy in secondary English. *Set: Research Information for Teachers,* (3), 51–57. https://doi.org/10.18296/set.0093

Morgan, J. (2012). *Teaching secondary geography as if the planet matters.* Abingdon: Routledge.

Murray, A. L. (1992). *Translations from the natural world.* Manchester: Carcanet Press.

Ranapiri, E., & Hawkes, R. (Eds.). (2022). *No other place to stand: An anthology of climate change poetry from Aotearoa New Zealand.* Auckland: University Press.

Simons, J. (2002). *Animal rights and the politics of literary representation.* Basingstoke: Palgrave.

Steiner, R. (1919). Supersensible knowledge and social pedagogical life-force, [Lecture, Stuttgart, 1919]. Retrieved May 30, 2024, from https://www.waldorflibrary.org/journals/22-research-bulletin/1194-springsummer-2012-volume-17-1-spirit-is-never-without-matter-matter-never-without-spirit-a-narrative-examination-of-a-college-of-teachers

Ward, C. (2005). Kanuka and manuka. In *Great Barrier Island Environmental News.* [website] Retrieved March 25, 2024, from http://www.greatbarrierenvironews.nz/Newsletters/Issue22/Kanuka%20Manuka.htm

White, D., Rudy, A., & Gareau, B. (2015). *Environments, natures and social theory: Towards a critical hybridity.* London: Palgrave Macmillan.

Wolfreys, J. (Ed.). (2015). *Introducing criticism in the 21st century.* Edinburgh: Edinburgh University Press.

Wordsworth, W. (1984). Lines written a few miles above Tintern Abbey on revisiting the banks of the Wye during a tour, July 13, 1798. In S. Gill (Ed.), *William Wordsworth* (pp. 131–135). Oxford: Oxford University Press.

2 Fostering environmental justice, climate justice and collective action through children's and young adult literature

Clare Echterling

Introduction

Previously, I have argued that American children's books about the environment overwhelmingly "tend to oversimplify and depoliticize" environmental issues and activism (Echterling, 2016, p. 286). From canonical works like *The Lorax* (Dr. Seuss, 1971/2012) to more contemporary titles like *The Magic School Bus and the Climate Challenge* (Cole & Degen, 2010), many of the books used in environmental education reduce ecological crises into simple, isolated problems and environmentalism into a set of individual consumer and lifestyle choices. Environmental issues of all kinds are deeply intertwined with systemic social problems like racial and economic inequality, but such texts overlook and obfuscate these connections. As activists and scholars have shown time and time again, marginalised peoples worldwide bear the least responsibility for climate change and other environmental crises, yet disproportionately bear the worst of their effects. Rather than address the intersections of environmental problems and social inequities in age-appropriate ways, many green books for young people altogether avoid issues of environmental injustice or the actors, policies and ideologies that drive it. To quote Nöel Sturgeon, such books fail to demonstrate how "social equality and environmental sustainability are interconnected" and instead "naturalize social inequality and disconnect environmental problems from their corporate causes" (2009, p. 103). Furthermore, books of this nature typically attribute environmental degradation flatly to "humanity" or "human activity", thus falsely suggesting that people everywhere are just as equally to blame as they are affected (Echterling, 2016, p. 290).

In the same vein, too many green children's and young adult (YA) books rehearse techno-optimist and neoliberal approaches to environmentalism. This is unsurprising given the dominance of neoliberalism in the United States, the United Kingdom, and other Western nations. However, technology and personal sacrifices alone are not enough to save the planet; we need green governance (Lukacs, 2017; Jensen, 2009; Levermann, 2019; Mann, 2019; Ribeiro & Soromenho-Marques, 2022; Ross, 2016). Yet many youth books fail to teach this lesson. They imply that systemic environmental problems can be solved by

DOI: 10.4324/9781032634524-4

technological innovation, market solutions and individuals who simply adopt eco-friendly lifestyles, rather than alongside democratic collective action, policy and legislation. Consequently, such books portray a reductive, exclusionary and ultimately futile type of environmentalism based on the politics of personal responsibility. They place the burden on youth and their families to save the planet at home by taking shorter showers or practising Meatless Mondays, but not through civic engagement and public forms of collective activism that can inspire widespread change. In this way, these books perpetuate the harmful myth that individual actions alone are enough to solve the environmental crisis. Furthermore, by over-emphasising or only portraying at-home environmentalism, they limit young people's political agency and misrepresent the role they can and are playing in the global environmental movement.

This is still the case for most green children's and YA books in the global north. However, there is a growing body of youth literature that foregrounds environmental and climate justice, the vital importance of collective action and the power of youth advocacy in public spheres. In part, this emergent canon speaks to an increased awareness of justice, equity and inclusion in environmental conversations. It also reflects the central role youth-led environmental movements have played since the late 2010s. Children and young people have always participated in environmental and climate advocacy, but never more visibly and in so many numbers since the global youth climate movement took off in late 2018 and early 2019. During that time, millions of young people joined Greta Thunberg's *Skolstrejk för klimatet* – creating the worldwide movement Fridays for Future – and began forming and organising with other grassroots youth coalitions, including the Sunrise Movement, Extinction Rebellion and Zero Hour. Today, youth are at the vanguard of the environmental movement. The publishing industry is responding in turn, releasing a welcome spate of new books about youth activists and activism.

This new wave of eco-literature is valuable for environmental educators who are looking to centre social justice and support young activists, which, this chapter argues, we must. More specifically, environmental justice literature can help students recognise the connections between environmental degradation and social inequities in the past and present. In doing so, these books invite critical thinking and conversation about environmental and climate justice in local and global contexts. Furthermore, by portraying diverse activists of all ages – from child, to teen, to adult, to elder – working together in a variety of ways and arenas to create systemic change, these texts model the many ways one can be a changemaker and illustrate the power of collective movements. Consequently, they correct the myth that environmentalism is simply about recycling, driving an electric vehicle or other green behaviours that may be unattainable or unrealistic for many students and families. Instead, they promote collaboration, solidarity and democratic public action, and invite readers to envision how they might work alongside others in the struggle for environmental justice and a safe climate for all. In what follows, I argue that environmental education must foster environmental justice and inclusive, collective forms of environmental activism and that teaching environmental justice literature can help us do so. I then propose a set of criteria for educators to evaluate potential book selections before discussing a handful of children's and YA books that meet those criteria.

Environmental justice, climate justice and environmental education

The environmental justice movement coalesced in the later decades of the 20th century. Unlike mainstream environmentalism, which historically has been indifferent and party to social injustice and violence, environmental justice focuses on the connections between ecological issues and social inequalities rooted in various aspects of identity, including class, disability, ethnicity, gender, race, nationality and sexual orientation. In 1991, delegates at the first People of Color Environmental Leadership Summit wrote and adopted the 17 *Principles of Environmental Justice*, which has since become a guiding document for the movement. As its eponymous name suggests, the movement is rooted in the concept of *environmental justice*, which holds that all people, regardless of background or social status, have the right to a healthy and safe environment and to *environmental self-determination* (Principles of Environmental Justice, 1991). The movement thus works to redress the disproportionate environmental devastation, deprivations and dispossession borne by marginalised and colonised communities around the world.

The climate justice movement developed as an offshoot of environmental justice in the late 20th and early 21st centuries as the threat of global warming became increasingly apparent. In 2002, a coalition of international groups, expanding on the *Principles of Environmental Justice*, created the *Bali Principles of Climate Justice* (Bali, 2002) to guide the growing movement. Like environmental justice, climate justice connects the climate crisis to social inequities and oppressive power structures. It recognises that climate change harms marginalised, underserved and primarily Black, Indigenous and people of colour communities around the world much more than those who are most responsible for it: corporations, nations and wealthy, predominantly white citizens in the global north.

These movements show that environmental solutions must be grounded in social justice, human rights and anti-racist democratic principles. If not, they risk sustaining or worsening inequality and human suffering. For the same reasons, environmental education must be based on the principles of environmental and climate justice. As Randolph Haluza-Delay argues, "environmental justice should be central to environmental education theory, practice, and research" (2013, p. 400). At best, environmental education that overlooks or merely pays lip service to social justice, diversity and inclusion will misinform and misguide students, providing an incomplete picture and encouraging them to practise environmentalist habits that can only serve as "an ineffectual band-aid on the wounds of the earth and its inhabitants" (Haluza-DeLay, 2013, p. 394). Daniel Kruidenier and Scott Morrison argue similarly (2013).

At worst, environmental education that fails to centre justice can inadvertently perpetuate or leave students vulnerable to myths that fuel exclusionary, illiberal and even violent strains of environmentalism such as the growing movement called *eco-fascism*.

Scholars debate the definition of eco-fascism, but the term now commonly describes the environmentalism of white supremacy and ultra-nationalism, which "purports to champion the protection of the planet via racist fear mongering" (Sipe, 2023, p. 4). In other words, its followers use environmental concerns "as a veneer for racist beliefs" (Rockey, 2021, *The origins of eco-fascism* section). More specifically, eco-fascists erroneously blame environmental decay on immigration and overpopulation, arguing that white people are being replaced by people of colour who are destroying the earth. They consequently advocate for brutal immigration restrictions, eugenic population control and genocide. Terrifyingly, eco-fascism has motivated several recent mass shooters, including the white nationalists who respectively attacked the Al Noor Mosque in Christchurch, New Zealand in 2019; a Walmart in El Paso, Texas in 2019; and a Tops supermarket in Buffalo, New York in 2022. While the movement is increasingly associated with far-right extremism and terrorism, eco-fascist ideas have purchase across the political spectrum and even mainstream appeal (Anson, 2020; Sipe, 2023). They are insidious and pervasive (Rockey, 2021; Sipe, 2023). For example, the Malthusian myth that overpopulation is a key cause of climate change and other ecological emergencies appears all-too-frequently in environmental discourse, including contemporary green children's and YA books such as Ter Horst and Peters, (2021). Centring social justice and liberation in environmental education can help educators and students alike resist and dismantle exclusionary and oppressive environmental ideas, practices and policies – not just those that are explicitly eco-fascist. At the same time, educating about environmental and climate justice can empower youth to envision and work towards a just and sustainable future for all.

Educating for collective action

Creating that future requires far more than recycling or growing vegetables at home. Eco-friendly habits are fruitless without systemic change, including the transition to a fossil-free economy (Lukacs, 2017). Yet many children's and YA eco-books and media exclusively or primarily promote individual solutions to ecological problems. Likewise, Kruidenier and Morrison argue, teachers that blame environmental crises on individuals and encourage students to save the planet through personal behaviours alone do them a disservice. They write:

> By not drawing attention to other sources of environmental degradation – from governments to militaries to corporations – and by not tending to a fuller depiction of citizen action, teachers leave students to fight for justice and the planet with credit cards at the mall. This can leave students feeling guilty, powerless, and no better prepared for the civic task that lies ahead.
> (2013, p. 432)

Indeed, the overemphasis on individual responsibility and solutions in environmental discourse is unfair and potentially harmful. A growing body of scholarship reveals that young people today feel overwhelmed, overburdened and hopeless in the face of ever-intensifying ecological disasters. These feelings and others comprise what mental health professionals are calling *eco-anxiety*, *climate anxiety* and *solastalgia* among other terms (Barry, 2022; Hickman, Marks, Pihkala, Clayton, Lewandowski, Mayall ...& van Susteren, 2021a, 2021b; Jaquette Ray, 2020b). The myth that individuals are personally responsible for causing and resolving the ecological crisis may fuel eco-guilt and anxiety that forecloses meaningful action (Jaquette Ray, 2020a, 2021).

To be fair, focusing on individual responsibility and solutions in the classroom or other educational spaces is understandable. Green behaviours like using reusable shopping bags and preserving water are relatively uncontroversial. So too is attributing environmental degradation to human activity. This focus allows educators to avoid delving into controversial topics and encouraging activist practices to which parents and governing bodies may object. As Rachel Hava Gordon observes, "teacher support of student movements [is] very tricky" and restrictions frequently limit educators' freedom to encourage activism (2009, pp. 205–206). I do not wish to underestimate or make light of the challenges teachers face in the contemporary political climate, but students deserve to learn the truth about environmental crises and the collective response they require. Young people today stand to inherit a scarred and poisoned planet. We must help them become "present and future citizens capable of acting on a societal as well as a personal level" (Jensen & Schnack, 1997, p. 164). As the global youth climate movement shows, many already are, but they need intergenerational allies to stand with and support them.

The notion that youth can or perhaps should only practise at-home environmentalism stems from the model of childhood which figures them as future citizens who cannot yet take part in political processes. Hava Gordon puts it well: "Youth are socially constructed as citizen participants only in the future tense: ill-equipped to participate in social and political decision making as youth, only capable of this participation as adults" (2009, p. 9). Hava Gordon and others, including Ben Kirshner, reject this model. Kirshner explains: "youth should not be treated as 'citizens of the future'; they are citizens now who experience, interpret, and sometimes resist the policies that organise their everyday lives, as enshrined in the UN Convention on the Rights of the Child" (2015, p. 5). Indeed, environmental education needs to recognise young people as citizens and empower them to participate as they choose in democratic processes and collective environmental practices, not just eco-friendly personal habits.

Doing so can prepare youth to take part in democracy and work for environmental justice throughout their lives. For Kathy Hytten, teaching students about historical and contemporary social movements and supporting their participation in them is essential to renewing democracy because it primes them for lifelong civic engagement and activism. Drawing on John Dewey's ideas in *Democracy and Education* (1916), Hytten argues that by learning about and engaging in social movements, students can develop "some of the most important habits of democratic citizenship", including solidarity and community organising (2016, p. 982). At the same time, this can help students "embrace the idea of democracy as an active way of life, rather than a static inheritance" (p. 994).

Teaching the histories and stories of environmental justice coalitions – including their large and small successes, setbacks and losses – can furthermore habituate youth to the slow, collaborative and unspectacular work of socio-environmental change. In a culture that idolises solitary heroes, it can also help dispel the sense that only exceptional individuals and their spectacular feats can make a difference. Instead, stories of environmental justice collectives can illustrate the power of ordinary people taking small steps together to inspire change, which most often occurs incrementally and in subtle ways. As the famous quote attributed to Margaret Mead reminds, we must "never doubt that a small group of thoughtful, committed citizens can change the world. Indeed, it is the only thing that ever has" (Mead, cited in Lutkehaus, 2009, p. 34). This tenet must be at the core of our approach to educating for environmental justice action.

Literature for environmental and climate justice

Youth literature about environmental justice and collectivity is a valuable tool in this project. Across genres and forms, literature can facilitate transformative reflection and dialogue about environmental and climate justice, liberated futures and collective action. Importantly, works that portray diverse youth-led and intergenerational coalitions and activism do more than just raise awareness; rather, they can model inclusive forms of advocacy and invite young readers to consider how they might stand up for environmental justice (Fletcher & Holyoke, 2023).

When choosing works to foster environmental and climate justice, educators may look for texts that fulfil the following criteria in age-appropriate ways. One, they should embody the movements' core concepts and guiding tenets, including those adopted in *Principles for Environmental Justice* and the *Bali Principles of Climate Justice*. Two, they should portray diverse and inclusive environmental justice movements and activists in the global south and north. Three, they should emphasise the importance of systemic change, solidarity and collective action over idealised solitary heroes, individual responsibility and green-washed lifestyle habits. In the next subsections, I highlight a small selection of exceptional non-fiction books that meet these criteria.

Selections for tweens, teens and young adults

The titles offered here can serve as foundational texts in environmental justice curricula for older children, teenagers and young adults. They explain key concepts and provide students with real-world examples of environmental injustice, activists and collective movements. They also provide practical advice on activism and may thus serve as guidebooks for student movements. Furthermore, they can contextualise students' interpretations of other environmental texts, helping them adopt an environmental justice lens through which to analyse fiction, poetry, drama, film and other media.

The Intersectional Environmentalist: How to Dismantle Systems of Oppression to Protect People + Planet

Leah Thomas's *The Intersectional Environmentalist: How to Dismantle Systems of Oppression to Protect People + Planet* (2022) provides an engaging and clear introduction to environmental justice. Thomas is a Black Gen Z environmental educator who founded the Intersectional Environmentalist, a youth-focused collective dedicated to making the environmental movement more inclusive and equitable. She coined the term *intersectional environmentalism* and launched the movement amid the Black Lives Matter protests that followed the murder of George Floyd in 2020. In a viral social media post, Thomas called for an anti-racist environmentalism that centres people of colour's voices and experiences, and invited people to take the Intersectional Environmentalist Pledge (Thomas, 2020). The term references Kimberlé Williams Crenshaw's theory of *intersectionality*, which analyses the ways different forms of discrimination such as racism, classism and sexism overlap and combine to impact people's lives in compounding ways.

Thomas's book situates intersectional environmentalism within the traditions of environmental justice, civil rights, Black feminism and *ecofeminism* (a philosophy and movement that focuses on the patriarchal oppression of women and the environment). By tracing intersectional environmentalism's lineage, Thomas helps illustrate that social movements never spring from nowhere; they are always indebted to those that came before. The chapter on environmental justice, furthermore, provides a history of the movement and introduces its principles alongside discussion questions for students and educators. From there, Thomas focuses on contemporary examples of environmental injustice primarily, but not exclusively, in the United States. These examples illustrate how structures of power, privilege and prejudice shelter or expose people to environmental harms.

Thomas intersperses praxis-focused sections throughout the book. These aim to help readers live out the Intersectional Environmentalist Pledge's tenets. For example, the chapter on privilege invites readers to consider how their own identities shape their experiences of the environment and activism. It then offers discussion questions and guidelines to help readers "use [their] privilege to advocate for" marginalised communities and "amplify the messages of ... environmental leaders" from those communities, including communities of colour, Queer communities and Disability communities (2022,

p. 80). Thomas also models these principles by incorporating insights and guidance from diverse environmental leaders throughout the book's praxis-focused sections and in the penultimate chapter, a "Toolkit" for intersectional environmentalists. While she does discuss eco-friendly lifestyle choices, Thomas emphasises their limitations and prioritises forms of outreach and collective activism that work toward systemic change.

How to Change Everything: The Young Human's Guide to Protecting the Planet and Each Other

For climate change curricula, Naomi Klein and Rebecca Stefoff's *How to Change Everything: The Young Human's Guide to Protecting the Planet and Each Other* (2021) offers a comprehensive overview of climate science, the history of the climate crisis and the climate justice movement for older children to young adults. Klein and Stefoff centre climate justice, using diverse examples of environmental injustice around the world to show that "not everyone experiences the effects of climate change to the same degree" (2021, p. 57). At the same time, Klein and Stefoff also tackle eco-fascism, encouraging readers to reject eco-fascist beliefs and "climate cruelty", "responses to climate disruption [that] are clearly *un*fair" (p. 92, 93). Rather, they argue, climate solutions must prioritise "justice and fairness", and they discuss what that might look like (p. 92).

Klein and Stefoff also dismantle the idea that everyone is equally responsible for climate change. Instead of blaming "humanity", they explain "that the climate crisis was overwhelmingly created by the world's wealthiest countries" at the expense of colonised people and land (p. 94). The chapters in Part Two: How We Got Here describe how this happened, tracing the history of the crisis from the Industrial Revolution and the beginnings of petro-capitalism to the present day. Together, these chapters show how fossil-fuel companies and neoliberal governments "that valued markets and profits over people and the planet" created the climate crisis (p. 158).

At its heart, the book is a call to action and a guide for young people to join the climate movement because "individuals alone cannot bring about the sweeping changes we need" (p. 56). On the contrary, Klein and Stefoff argue, "only social movements can turn the tide of climate change" (p. 162). They provide ample examples of how youth-led and intergenerational movements around the world are working to "turn the tide", including the Heiltsuk Nation's successful campaign against the proposed Northern Gateway pipeline in British Columbia (pp. 163–171). These diverse examples model various types of climate action, thus offering readers inspiration and manifold "path[s] into activism" (p. 258). Like *The Intersectional Environmentalist*, *How to Change Everything* also provides a "toolkit" chapter that includes ideas and practical guidance for youth working with peers and adults to create change.

By youth, for youth: books by Young environmental activists

As young environmental justice leaders including Zanagee Artis, Greta Thunberg and Autumn Peltier have gained prominence, some have begun writing and editing books to inspire other readers to join the movement. There are too many to discuss each here, but these books are useful companions to foundational texts like *The Intersectional Environmentalist* and *How to Change Everything* in environmental justice curricula. These manifestos allow young activists to tell their own stories and appeal directly to other young people while pushing back against adult supremacy and the idea that youth have no place in politics. At the same time, their stories show how ordinary kids and teenagers can become agents of change, thus mapping out some of the many ways readers can become changemakers too.

For example, in *Youth to Power: Your Voice and How to Use It* (2020), Zero Hour co-founder Jamie Margolin rejects the myths that a person must be exceptional to become a changemaker and that youth cannot make a difference. Instead, she stresses how ordinary young people allying together and with adults can inspire change. "You", she declares to the reader, "belong in the political world" and "with us, the young people changing the world" (p. xvi). Throughout the book, she offers her experiences and insight to guide readers on their journeys to becoming "scrappy activists" (p. xvi, xvii).

Similarly, in *Imaginary Borders* (2020), Xiuhtezcatl Martinez relates his diverse experiences as a youth environmental activist to mobilise his readers. Martinez, who in part uses hip hop to raise awareness and inspire action, emphasises the importance of creativity and pleasure in activist work to show that it can and should take on many forms. In doing so, he – like Margolin – argues that there is a place for every young person in the environmental movement. These themes also become clear through *Tomorrow Is Too Late: An International Youth Manifesto for Climate Justice* (2021), an essay collection compiled by the teen climate activist Grace Maddrell.

Maddrell's collection features essays by youth activists from around the world, including Asam Sbaih (Palestine), Devansh Desai (India), Maureen Damen (Senegal) and Rita Naumenko (Russia). Consequently, it amplifies the voices of campaigners whose homelands, perspectives and contributions mainstream Western media and environmental discourse routinely overlook. The news and other media overwhelmingly focus on white and global northern youth activists – especially Greta Thunberg, who is regularly figured as an idol and solitary superhero (despite her wishes). As Katja Schreiber points out, this bias extends to youth environmental literature and media, which predominantly features the stories and voices of Thunberg and other activists from wealthy Western nations, but typically overlooks and silences global southern and African activists (Schreiber, 2021).

The Ugandan activist Vanessa Nakate's movement memoir, *A Bigger Picture: My Fight to Bring a New African Voice to the Climate Crisis* (2021), confronts this problem clearly and can thus help environmental educators address it. In the memoir, Nakate recounts the time when the Associated Press cropped her – the only Black person and the only person of colour – out of a photo of climate delegates at the 2020 World Economic Forum. As Nakate explains, the incident signified how Ugandans, Africans and other global south citizens have been excluded from climate conversations and decision-making. In *A Bigger Picture*, consequently, she calls for a climate justice movement that uplifts diverse leaders and centres Africa, the global south and the communities most affected by climate disruption.

Selections for younger students

Books for environmental justice

Written by Catalina M. de Onís, Hilda Lloréns, Mabette Colón Pérez, and Khalil G. García Lloréns, and illustrated by Colón Pérez, the bilingual picturebook *La Justicia Ambiental es Para Ti Y Para Mí / Environmental Justice Is for You and Me* (2021) offers a clear introduction to environmental justice, community organising and collective power for young children and beginning readers. The book begins by defining key concepts. To illustrate them, it then tells Colón Pérez's story of how, as a child, she and her father worked with their community to fight coal ash pollution created by a local power plant in Guayama, Puerto Rico. Throughout, the text gently encourages the child reader to work with their caretakers and communities to combat environmental injustice. Due to the clarity with which it explains environmental justice's basic tenets, it can prepare students to engage critically with other books, including those that illuminate (neo)colonialism's violence against people and the planet.

As the *Principles of Environmental Justice* hold, true environmental justice requires the liberation of all peoples from "colonisation and oppression". The movement is thus inseparable from the struggle to decolonise Indigenous lands and communities and secure the rights of all people to "sovereignty and self-determination" (*Principles,* 1991, no. 11). This fundamental shines through in Patricia Gualinga, Laura Resau and illustrator Vanessa Jaramillo's *Stand as Tall as the Trees: How an Amazonian Community Protected the Rain Forest* (2023). Beautifully written and illustrated, this autobiographical picture book details Patricia (Paty) Gualinga and the Kichwa (alternatively Quichua) people's fight to save their ancestral homes and ways of life after the Ecuadorian government sold their lands to oil companies. Although focused through Paty's first-person narrative, the book emphasises how the Kichwa people have worked together with other Indigenous groups "to protect their culture, well-being, and sacred forest" and invites readers "to join them" (p. 29). Importantly, the afterword also contextualises the Kichwa campaign among other Indigenous movements around the world, showing readers that their struggles are not unique. Rather, it illustrates that Indigenous peoples are on the frontlines of environmental justice, "standing up for ancestral lands" and cultures in the face of ongoing colonial and neocolonial onslaughts (p. 30).

Environmental justice is also linked to workers' rights, which two recent picture books about the Brazilian labour organiser and environmentalist Chico Mendes make clear. *The Life of / La vida de Chico* (Rodriguez, Stein & Reyes, 2022) and *Forest Fighter: The Story of Chico Mendes* (Ganeri & Carpentier, 2022) both invite conversation and reflection about the exploitation of workers, poor communities and the land. Written by Patty Rodriguez and Ariana Stein, with art by Citlali Reyes, *The Life of / La vida de Chico* is a bilingual board book for infants and young children. Written by Anita Ganeri and illustrated by Margaux Carpenter, *Forest Fighter: The Story of Chico Mendes* (Ganeri & Carpentier, 2022) is for readers aged seven to twelve. In both, the text and images work synchronously to tell how Mendes and his rural rubber trapper community organised first against the rubber barons and then the cattle ranchers who forced them out of their homes and cleared the rainforest to create pastureland. They each celebrate Mendes' legacy without suggesting he ever worked alone. Rather, they focus on the strength of collective movements and the ongoing fight to protect people, animals and the planet together.

Books for climate justice

The picture books described above can also facilitate conversations about climate justice, but elementary climate curricula benefit from books specifically on the subject. The number of kids' climate books has risen dramatically in recent years, but few truly centre on climate justice and collective responses to the climate crisis while encouraging children and their caretakers to take action beyond the home. Here, I recommend three that do: *We Have a Dream* (2021), *No World Too Big: Young People Fighting Global Climate Change* (2023) and *Count on Us! Climate Activists from One to a Billion* (2022).

Written by the British-Bangladeshi youth environmental justice activist Mya-Rose Craig and illustrated by Sabrena Khadija, *We Have a Dream* (Craig & Khadija, 2021) is appropriate for both child and young adult audiences. In it, Craig profiles 30 young Black, Indigenous and people of colour environmentalists to uplift their voices and raise awareness about their contributions to the movement. Based on interviews she conducted during the COVID-19 lockdown, the profiles combine into a manifesto for climate justice. Furthermore, they showcase many types of environmental advocacy, offering inspiration and guidance for aspiring activists.

Similarly, *No World Too Big* (2021) uses poems and short blocks of prose to introduce a handful of youth climate leaders and collectives. Edited by Lindsay H. Metcalf, Keila V. Dawson and Jeannette Bradley (also illustrator) and aimed at readers aged five to nine, the book's contributors use various poetic forms to educate readers about the 2015 Paris Agreement, climate justice and youth activists, including Zanagee Artis, the Marshall Islands Youth Leadership Coalition and Marinel Ubaldo. Concrete action steps accompany many of the poems, prompting readers to practise creative forms of climate advocacy, such as "stag[ing] a play about the effects of climate change" (p. 16).

The poems also provide opportunities to teach students – even ones older than the book's target audience – about verse forms, literary terms and the role of climate literature.

For slightly younger audiences, Gabi Snyder and illustrator Sarah Walsh's *Count on Us! Climate Activists from One to a Billion* (2022) is an alphabet and counting book that introduces climate activism and the power of collective action. Through complementary text and illustrations featuring diverse environmentalists of all ages, the book subtly promotes climate justice and encourages children to join others in "zooming towards a better future" (p. 24). Although Snyder does suggest individual actions the reader and their adults might take, the narrative also promotes collective, public-facing action and invites conversation about various forms of activism, including community organising and public demonstrations.

Conclusion

The Principles of Environmental Justice and the *Bali Principles of Climate Justice* both mandate the "education of present and future generations" in ways that adhere to the movements' core tenets and goals (*Bali,* 2002, no. 25; *Principles,* 1991, no. 16). The Bali principles furthermore hold that future generations have the right to "a stable climate and healthy planet" (no. 27). Yet the students of today and tomorrow are facing a future defined by ever-worsening environmental disasters and deepening social inequalities. They are standing up in unprecedented numbers to protect their futures, but they cannot do so alone – they need adults to stand with them in solidarity. As environmental educators, we must do our part by orienting our pedagogies towards environmental and climate justice and educating for collective action. We can do this, in part, by reaching for children's and YA literature that moves beyond exclusionary and futile types of environmentalism and instead points towards a just and sustainable future.

References

Anson, A. (2020, October 21). No one is a virus: On American ecofascism. EHN Blog. *Environmental History Now.* https://envhistnow.com/2020/10/21/no-one-is-the-virus-on-american-ecofascism/

Bali Principles of Climate Justice. (2002). *Ejnet.org.* www.ejnet.org/ej/bali.pdf

Barry, E. (2022, February 6). Climate change enters the therapy room. *The New York Times.* https://www.nytimes.com/2022/02/06/health/climate-anxiety-therapy.html

Cole, J., & Degen, B. (2010). *The magic school bus and the climate challenge.* London and New York: Scholastic.

Craig, M. R., & Khadija, S. (2021). *We have a dream.* London: Magic Cat Publishing.

de Onís, C., Lloréns, H., Colón Pérez, M., & García-Lloréns, K. (2021). *La justicia ambiental es para ti y para mí / Environmental justice is for you and me.* Cabo Rojo, Puerto Rico: Editora Educación Emergente.

Dewey, J. (1916). *Democracy and education: An introduction to the philosophy of education.* New York, NY: Macmillan.

Echterling, C. (2016). How to save the world and other lessons from children's environmental literature. *Children's Literature in Education, 47,* 283–299. https://doi.org/10.1007/s10583-016-9290-6

Fletcher, L., & Holyoke, E. (2023). Reading the word and the world: Portrayals of activism in children's literature. *The Reading Teacher, 76*(6), 713–723. https://doi.org/10.1002/trtr.2191

Ganeri, A., & Carpentier, M. (2022). *Forest fighter: The story of Chico Mendes.* Northampton, MA: Crocodile Books.

Gualinga, P., Resau, L., & Jaramillo, V. (2023). *Stand as tall as the trees: How an Amazonian community protected the rain forest.* Watertown, MA: Charlesbridge Publishing.

Haluza-DeLay, R. (2013). Educating for environmental justice. In M. Brody, J. Dillon, R. Stevenson, & A. Wals (Eds.), *International handbook on environmental education research* (pp. 394–402). New York: Routledge.

Hava Gordon, R. (2009). *We fight to win: Inequality and the politics of youth activism.* New Brunswick, NJ: Rutgers University Press.

Hickman, C., Marks, E., Pihkala, P., Clayton, S., Lewandowski, R. E., Mayall, E. E. … van Susteren, L. (2021a). Climate anxiety in children and young people and their beliefs about government responses to climate change: A global survey. *Lancet Planetary Health, 5*(12), 863–873. http://www.thelancet.com/planetary-health

Hickman, C., Marks, E., Pihkala, P., Clayton, S., Lewandowski, R. E., Mayall, E. E. … van Susteren, L. (2021b). Young people's voices on climate anxiety, government betrayal and moral injury: A global phenomenon. *Preprints with the Lancet.* http://dx.doi.org/10.2139/ssrn.3918955

Hytten, K. (2016). Globalization, democracy, and social movements: The educational potential of activism. *Educational Philosophy and Theory, 48*(10), 981–996. http://10.1080/00131857.2016.1150801

Jaquette Ray, S. (2020a). *A field guide to climate anxiety: Keeping your cool on a warming planet.* Oakland, CA: University of California Press.

Jaquette Ray, S. (2020b, February 11). OK, doomers! The climate generation has arrived. *Edge Effects.* Retrieved February 29, 2024, from https://edgeeffects.net/climate-generation/

Jaquette Ray, S. (2021, June 29). Is climate anxiety bad for the planet? *Los Angeles Times.* https://www.latimes.com/opinion/story/2021-06-29/climate-anxiety-psychology-community-action

Jensen, B. B., & Schnack, K. (1997). The action competence approach in environmental education. *Environmental Education Research, 3*(2), 163–178. https://doi.org/10.1080/1350462970030205

Jensen, D. (2009). Forget shorter showers: Why personal change does not equal political change. *Orion.* https://orionmagazine.org/article/forget-shorter-showers/

Kirshner, B. (2015). *Youth activism in an era of education inequality.* New York, NY: New York University Press.

Klein, N., & Stefoff, R. (2021). *How to change everything: The young human's guide to protecting the planet and each other.* New York, NY: Atheneum Books for Young Readers.

Kruidenier, D., & Morrison, S. (2013). Avoiding the banking model in social and environmental justice education: Interrogate the tensions. *Educational Studies, 49*, 430–442.

Levermann, A. (2019, July 10). Individuals can't solve the climate crisis. Governments need to step up. *The Guardian*. https://www.theguardian.com/commentisfree/2019/jul/10/individuals-climate-crisis-government-planet-priority

Lukacs, M. (2017, July 17). Neoliberalism has conned us into fighting climate change as individuals. *The Guardian*. https://www.theguardian.com/environment/true-north/2017/jul/17/neoliberalism-has-conned-us-into-fighting-climate-change-as-individuals

Lutkehaus, N. (2009). Margaret Mead: Public anthropologist. *Anthropology Now, 1*(1), 29–35.

Maddrell, G. (Ed.). (2021). *Tomorrow is too late: An international youth manifesto for climate justice*. London: The Indigo Press.

Mann, M. E. (2019). Lifestyle changes aren't enough to save the planet. Here's what could. *Time*. https://time.com/5669071/lifestyle-changes-climate-change/

Margolin, J. (2020). *Youth to power: Your voice and how to use it*. New York, NY: Hachette Book Group.

Martinez, X. (2020). *Imaginary borders*. New York, NY: Penguin Workshop.

Metcalf, L. H., Dawson, K. V., & Bradley, J. (Eds.). (2021). *No world too big: Young people fighting global climate change*. Watertown, MA: Charlesbridge Publishing.

Nakate, V. (2021). *A bigger picture: My fight to bring a new African voice to the climate crisis*. Boston, MA: Mariner Books.

Principles of Environmental Justice. (1991). *Ejnet.org*. www.ejnet.org/ej/principles.pdf

Ribeiro, S., & Soromenho-Marques, V. (2022). The techno-optimists of climate change: Science communication or technowashing? *Societies, 12*(2), 64. https://doi.org/10.3390/soc12020064

Rockey, F. (2021, March 21). The dangers of eco-fascism and why it's a 'veneer for racist beliefs.' *EuroNews*. https://www.euronews.com/green/2021/03/21/the-dangers-of-eco-fascism-and-why-it-s-a-veneer-for-racist-beliefs

Rodriguez, P., Stein, A., & Reyes, C. (2022). *The life of / La vida de Chico*. Los Angeles, CA: Little Libros.

Ross, A. (2016). Climate change. In J. Adamson, W. A. Gleason, & D. N. Pellow (Eds.), *Keywords for environmental studies*. New York, NY: New York University Press.

Schreiber, K. (2021). Ecocriticism and youth climate heroes in Southern Africa: Centering global south perspectives and activism in *Isaura*. *The Lion and the Unicorn, 45*, 211–225. https://muse.jhu.edu/article/819555

Seuss, Dr. (1971/2012). *The Lorax*. London: HarperCollins UK.

Sipe, W. J. (2023). *Resisting racist fantasies in online discourse: Eco-fascism & the dangers of capitalist alienation* (30566422). (Doctoral dissertation, University of Nebraska – Lincoln, NE). ProQuest Dissertations and Theses Global.

Snyder, G., & Walsh, S. (2022). *Count on us! Climate activists from one to a billion*. Concord, MA: Barefoot Books.

Sturgeon, N. (2009). *Environmentalism in popular culture: Gender, sexuality, and the politics of the natural*. Tucson, AZ: University of Arizona Press.

Ter Horst, M., & Peters, M. (2021). *Palm trees at the North Pole: The hot truth about climate change* (Laura Watkinson, Trans.). Vancouver, BC: Greystone Kids.

Thomas, L. (2022). *The intersectional environmentalist: How to dismantle systems of oppression to protect people + planet.* New York, NY: Voracious (Little, Brown and Company).

Thomas, L [@greengirlleah]. (2020, May 28). *Social justice cannot wait. It is not an optional "add-on" to environmentalism. It is unfair to opt in and* [Text-Based Graphic]. Instagram. https://www.instagram.com/p/CAvaxdRJRxu/?img_index=3

3 Humanising science through the arts

Martin Braund

Introduction

> Both the known and the unknown, the two worlds of our ancestors, nourish the human spirit. Their muses, science and the arts, whisper: Follow us, explore, find out.
>
> (Wilson, 1998, p. 233)

It has been commonplace in educational systems throughout the world to separate the arts from the sciences. While science and arts cannot be said to be the same, they share basic objectives shaped by human aspiration and endeavour, as Wilson suggests. While science searches for empirically validated truths about the world, the arts offer aesthetic interpretations of it (Hadzigeorgiou, 2016). While this seems to point to philosophical and educational separation, one could also take the view that all forms of knowledge are complementary, on the grounds that they contribute to a fuller and richer experience and understanding of the world. The arts project a human presence in the universe that science is often trying to avoid. In predicting the behaviour of phenomena and systems, scientists develop languages using models and equations that rarely include a human element. It is as if science has developed an increasingly privileged set of communication modes that take it further away from the everyday experiences of the mass of people trying to gain access to it. This is why I use this chapter to show how the arts (language arts) could be used to *humanise* science, capitalising on improving access through aesthetic and representational methods afforded by the arts.

I want to be clear from the start that, while I do not necessarily see the *arts* in general as entirely synonymous with *language arts*, I cannot exclude many art forms as ways of communicating about science and in Environmental Education. Indeed, art was the first human communication before the evolution of spoken languages (see the following section). Ison and Bramwell-Lalor (2023) see three domains of experience applied to Environmental Education: cognitive – knowledge of nature, psychomotor – actions with/in nature, affective – attitudes and emotions about nature. They see all three domains being accessed through art and design products and processes. Their list seems to

DOI: 10.4324/9781032634524-5

me to encompass most communication, and so language arts forms, that I can think of: creative writing, dance, drama, drawings, fabric, films, models, music, literature, natural samples (for example leaves, mud, rocks, soil and wood), painting, photographs, poetry, sculptures and videos (Ison & Bramwell-Lalor, 2023, p. 966). I also want to be clear that, while I write about the arts in relation to the sciences, most of what I have to say applies equally to Environmental Education, as will become clear through the specific personal narratives interwoven with my discussion.

Although the clashes between science and the arts in education and society, once described by C.P. Snow (1959) as *The Two Cultures*, might seem to continue into the 21st century, there has been increasing recent interest in what the arts might offer the sciences. A number of studies examine a deepening relationship between the arts and sciences, suggesting that the enterprise of science and of learning benefits from the arts (Braund & Reiss, 2019; Colucci-Gray, Burnard, Cooke, Davies, Gray & Trowsdale, 2017; Hadzigeorgiou, 2016; Tytler, Prain & Hannigan, 2020). In terms of research evidence to support beneficial outcomes from arts-science and Environmental Education links, there are studies showing stronger connections with nature (Arbuthnott & Sutter, 2019; Bruni, Winter, Schultz, Omoto & Tabanico, 2017), increased collegiality and collaboration (Mills, 2013), improved motivation and engagement in science subjects (Henriksen, 2014) and improved mental and physical development of learners (Dhanapal, Kanapathy & Mastan, 2014). There is also evidence that arts involvement can enhance performance in formal assessment in science (Root-Bernstein & Root-Bernstein, 2013; Ruppert, 2006).

In the main part of this chapter, I discuss arts in relation to science and Environmental Education in three themes: the languages of science, the question of science and arts *capital* and finally, considering the aesthetics and beauty of science. I have drawn on my personal experiences as teacher, educator and researcher to provide counterpoint and illustration to the ideas presented. Before discussion of each of these themes I provide a narrative looking at the evolving relationship between the arts and sciences and the development of Environmental Education, mainly as it has been experienced in the UK (specifically England and Wales).

An uneasy relationship? Arts, science and Environmental Education

Art seems to be as old as cognate human existence and predated communication by written or spoken languages. It is commonly recognised that art, in expressive visual form, dates as far back as the late Palaeolithic (about 40,000 BCE). With the development of language, new literary art forms of poetry, fiction and theatre became possible. In one of the earliest manifestations of an arts-science link, in Ancient Greek times, the *muse* of poetry was used to communicate ideas in astronomy (Braund & Reiss, 2019).

The idea of *science*, as separate from *natural philosophy*, only came into common use in the early 19th century (Heidegger & Grene, 1976). Stemming from Baconian beliefs of inductivism and empiricism, scientific *methods* were considered to be fundamental to modern (empirical) science. Thus, the enterprise of doing science has often been (and still is) presented as a progressive narrative in which true theories replaced false beliefs (Bereiter, 1994; Braund & Reiss, 2019). Alternative, expansive interpretations of science, such as those of Thomas Kuhn, portray science in more nuanced terms, as competing paradigms or conceptual systems in a wider matrix that includes intellectual, cultural, economic and political themes, considered by traditional "Baconists" to be outside science (Kuhn, 1970). Unfortunately, for many school students today, the more restricted view of science founded on Baconian traditions has often led to perceptions of science as a fact-driven enterprise divorced from the culture in which it exists and which it serves.

An early example of how Kuhnian paradigm shifts might be influenced by the arts comes from the story of Neils Bohr and his seminal work on the structure of the atom. Bohr needed a new way of conceptualising the atom to allow for the erratic behaviour of electrons, stepping beyond the classic planetary model of electrons orbiting a nucleus in a planar ellipse. It has been argued that his model of atomic structure was stimulated by contemporaneous changes in literature and the arts, such as Cubism (Clarke, 2014). Bohr maintained that the form electron paths took meant electrons were not like little planets at all. Instead, they were like one of Picasso or Braque's deconstructed pictures, a blur of brushstrokes that only made sense once you stared at it for long enough from different angles. The idea that art stimulates paradigm shifts continues in the 21st century. Structural patterns in DNA have been progressed by looking at Escher's art (Seeman, 2014). Seeman's work has also been aided by looking at patterns in Roman mosaic tile floors and Islamic art tiling at the Alhambra. While I do not claim that, without Cubism or the art used by Seeman, these scientific shifts would not have arisen, my point here is that the existence of new ways of looking at the world in the arts opens up spaces in which new thoughts about how the physical world works are more likely. Root-Bernstein and Root-Bernstein (2013) present the idea that the arts provide a sort of widened cognitive platform for scientific thinking, developing habits of thought and action that include practising, persevering and trial-and-error problem solving valued in science.

Environmental Education

At this point it is helpful to elucidate the emergence of Environmental Education as it has been variously conceived, most notably in England and Wales and with some lesser discussion of initiatives in the US. It seems that Environmental Education (EE from now on) arose as an identifiable educational entity, at least as far as England and Wales are concerned, in the 1960s, with roots in 18th- and 19th-century romanticism and school nature study

(Palmer, 1998; Van Boeckel, 2013). It should be noted from here on that the curriculum for England and Wales is operated separately from the other constituent parts of the United Kingdom (Scotland and Northern Ireland).

By the time of the introduction of a National Curriculum for schools in England and Wales in 1990, EE was positioned as a documented, though non-statutory, part of the curriculum (National Curriculum Council, 1990). EE was afforded the status of a *cross-curricular theme* along with other areas such as Health Education and Citizenship. In reality, EE was rarely taught as a separate subject, usually hived off into science subjects or as part of Geography. School inspectors merely asked how schools were addressing the cross-curricular themes rather than requiring evidence of learning outcomes as they did for science and other disciplinary areas, resulting in EE having low status. The conception of EE in the National Curriculum was as an education *through*, *for* and *about* the environment. The American environmental educator Steve Van Matre commented that this merely positioned EE as having vague aims:

> environmental education became a bit of everything to everyone, and consequently not much of anything to anyone. It became anything anybody wanted it to be at the moment.
>
> (Van Matre, 1999, p. 46)

Contemporaneous with the development of EE in England and Wales, there was a growing movement throughout North America for EE to be more emotionally responsive and tuned to individual environmental empathy and experience. Earth Education, established by Van Matre in 1974, aimed to foster learners to have feelings of security and comfortability in their own homes, but also for the planet itself, effectively an effort to acclimatise learners to the earth and its natural systems. Van Matre, along with other influential US educators such as Joseph Cornell (Cornell, 1979, 1989), devised sets of activities to acclimatise learners, tuning them into their locales. Earth Education reasons that, when one is too full of one's own thoughts, one cannot make room so easily for the impressions of the other life around oneself. Van Matre asserts that acclimatisation focuses on the feelings, on sharpening non-verbal skills like watching and waiting, silencing and stilling, opening and receiving (Van Matre cited in Van Boeckel, 2013, p. 117).

When teaching as a lecturer in Environmental Science and Environmental Education at Bretton Hall College in West Yorkshire in the 1990s, our team was heavily influenced by the Earth Education movement and the writings of Van Matre and Cornell. On one occasion, I led a group of about 20 postgraduate education students (specialists in Environmental Science) on a morning of Earth Education activities. The weather was warm and sunny with clear blue skies – a perfect day, one might say. It was early morning but with the heat building. I told the group to remove shoes and socks and lie in a circle under the dappled shade of a Cedar of Lebanon tree in the grounds of the college. Eyes closed and silent, we listened to birdsong and the rustling of branches, felt leaf needles and grass with our feet and toes, experienced the changing temperatures and light intensities as sunlight filtered through the tree's canopy. Later I asked individual students to find a special place in the grounds and to spend some time there, returning to describe in turn their place using whatever means they wanted: a sketch, some poetry or prose or just telling about the place. On another occasion in Finland, I visited a Nature School based on a small island not far from the Finnish capital, Helsinki. We were asked to make a jam jar aquarium, collecting representative parts of the shore ecosystem; some pebbles, sand, sea snails, seaweed and so on but also to make our product as beautiful as possible.

I relate these experiences to show how communication using visual and other language arts draws on humanised experiences. A critique of these sorts of approaches is that direct empathy experiences of nature, particularly in what some call "pristine or wilderness environments", can become the holy grail of EE or are used as a panacea for all EE learning, with demonstrable affective outcomes but little of cognitive value (Russell, 1999, p. 127). Van Matre counters these views, being adamant that activities should have a clear structure and that their purposes and intended outcomes are clearly stated and communicated to learners. He contrasts his aims sharply with what he regards as one of the prevalent myths in EE, that discovery learning is unstructured, "a kind of hippie grooving on nature" (Van Matre, cited in Van Boeckel, 2013, p. 117).

Arts for Environmental Education

In the 1940s Herbert Read saw art as a primary driver in education maintaining that without an art experience no one could properly open themselves up to the living earth (Read, 1946). Finnish artist Osmo Rauhala sees the crucial role of art as providing ways of interpreting and signifying experiences of the world, refining our perception and sensitivities for the mystery of things around us (Rauhala, 2003, p. 24). In this vein, and again in Finland, Meri-Helga Mantere has established an approach, Arts-based Environmental Education (AEE), where art is not an added quality, the icing on the cake, but rather the point of departure in the effort to find ways in which people can connect to their environment.

> as an environmentalist and as an art teacher … I try to support fresh perception, the nearby, personal enjoyment and pleasure of perceiving the world from the heart. To achieve that, it is necessary to stop, be quiet, have time and feel psychologically secure in order to perceive the unknown.
>
> (Mantere, 1998, p. 32)

In this quote we can see connections to the Earth Education movement – they have similar intentions for creating environmental empathy through arts and communication. This idea of arts providing an alternative to formalised scientific communication and languages is developed in my first theme.

The languages of science

A major problem for learners is that learning science requires informed use of multiple languages for understanding and representation. Everyday words, such as light, energy and force, have specific meanings in science that clash with meanings in normal speech and in everyday life. Energy is regarded in science as the capacity to do work, and that it is transferred from one form to another rather than used up. In everyday speech we might say, we just do not have *enough energy* to do something, opposite to the science view of conservation. Force might be used in the connotation of a police force or *being forced* to do something, rather than as a physical quantity to be measured in Newtons. Thus, science language use relies on what Wittgenstein called understanding the "rules of the game", where the language we use and how we apply it depends very much on knowledge and understanding of the specific language context one finds oneself in (Wittgenstein, 2010).

The idea of language rules also applies in how science uniquely uses symbolic and semiotic systems of representations making it often seem like learning a foreign language (Bleicher, Tobin & McRobbie, 2003). Science teachers communicate ideas with mathematical and chemical symbols, such as in equations and chemical formulae. These multiple languages of science create significant subject-specific barriers to student learning that do not occur in other school subjects (Tytler et al., 2020; Wellington & Osborne, 2001). Thus, I maintain that communication through a variety of visual, spoken and alternative written language modes (language arts as alternatives to written expository texts), drawn from the arts, offers science ways of breaking down the barriers that so many pupils find hard to cross. An example from my personal experience shows what can be achieved.

I was once engaged, as an advisory teacher, in helping primary school teachers feel more comfortable about teaching science. In the early 1990s, following the introduction of a National Curriculum in England and Wales, primary teachers were required to teach science having had no science in their initial training. One teacher of a class of ten-year-olds was concerned about ways of communicating science through the formal narrative of reporting investigations (method, results and conclusions). This was a natural consequence of her own experiences as a learner – falling back on the ways in which she had been taught. I conducted a lesson with her where groups of pupils pursued different investigations (cars rolling down a slope, dissolving sweets, swinging a pendulum, mopping up spills with paper towels). I asked groups to come up with whatever ways they could think of to report back their findings, giving some suggestions that almost anything goes (except the formal science ways of writing or telling us the method, results or conclusions). To my delight and this teacher's amazement, her pupils' creativity came to the fore. There were poems, rap singing, poster pictures and even a short acted-out playlet. I think this teacher went away a little happier that science could engage and be fun at the same time as following the basic processes enshrined in empirical science at pupil level. I use this example to point out that the arts (language arts) provide alternative narratives to the ones often assumed to represent science successfully and that many teachers adhere to.

Previously I discussed the contribution of the Arts for Environmental Education (AEE) movement in Finland in the wider context of EE. While some of AEE's activities are not confined to the visual arts, many of Mantere and Van Boeckel's examples use fine arts/painting as communication modes. While these often refer to generating artistic artefacts, there are cases in wider science education that use fine art pedagogy through art appreciation. An example in higher education shows that arts-based pedagogy in science is not just limited to schools. At the University of California, medical students used masterpieces by Van Gogh, Rembrandt, Kandinsky and da Vinci, to improve observation and pattern recognition skills in clinical situations (Shapiro, Rucker & Beck, 2006). Outcomes from following a fine arts programme developed better skills in emotional recognition, cultivation of empathy, identification of story and narrative and awareness of multiple perspectives. In this case, it seems that using fine art stimulated learning from multiple perspectives.

There are documented examples of fine art (painting and drawing) being used in expressive modes for school pupils to communicate scientific ideas in creative ways. In a workshop on "Hydrogen Fuel-Cell Cars", students discussed the basic workings of a fuel cell and then designed their own car of the future (Halpine, 2004). In a similar vein, pupils in a school in England were encouraged to design and draw an insectivorous plant showing its feeding and reproductive adaptations, after studying original letters by Charles Darwin describing his work on similar plants (Stafford, 2015, p. 43). Their pictures showed high levels of integration between artistic creativity and substantive scientific writing.

It is possible these art activities work in science lessons because they tap into visual thinking of learners. It is often said that today's school pupils inhabit a visual world dominated by television, games, computers, mobile phones and tablets, films and so on. Bob Horn believes visual thinking is far more effective at conveying convoluted ideas than conventional methods of communication (Horn cited in Gangwer, 2009, p. xi). It has been claimed that visual thinking translates into problem-solving ability. Visual thinkers literally see the answers to problems, enabling them to build entire information systems using their imaginations (Gangwer, 2009, p. xi).

In the last ten years or so, I have been engaged in researching and writing about drama to teach science, but my first real foray into this field as an academic came at Bretton Hall in the late 1990s (see Braund, 1999a). Bretton Hall was (until the early 2000s) an Arts and Education College of the University of Leeds. Many of my students came to me to learn curriculum science as teacher education students whose specialism was in arts subjects such as literature, fine arts and theatre and drama. Their fear of, and sometimes loathing for, science came from their poor experiences as school learners. My strategy was therefore to draw on arts pedagogy, particularly drama, to help them learn background concepts such as circuit electricity. Success was evidenced by my student teachers, who had been taught using drama, outperforming students in the same year group who had not had the benefit of these sessions when assessed using standardised tests of knowledge of electricity (Braund, 1999b).

I believe that drama activities, like those used at Bretton, work because they act as effective analogous learning. The languages of science, I discussed earlier, often portray abstract ideas conceptualising invisible components (energy, molecules, electrons, biological cell components and so on), and, as mentioned before, these ideas are communicated through symbolic and semiotic languages. This makes learning for pupils, who might already have a number of alternative views of how the world works, even more problematic. Rationalising between these two worlds, the science and the everyday, requires differentiation between and integration of two ways of explaining and seeing (Scott, Mortimer & Ametller, 2011). Integration of ideas requires accommodation of new ideas with those already held that provide more workable, rational and generalisable explanations of the world. To achieve integration means making abstract ideas and theories of science more plausible. In the science teacher's pedagogical toolkit, this is often done by the use of analogy and metaphor (Aubusson, Harrison & Ritchie, 2006). For example, analogues, supposedly from real-world experience such as the hot water system of a house or a ski lift, have been used to explain concepts of current, voltage and resistance in electrical circuits. The problem here is that the analogues themselves may not be fully understood by pupils, confer even more alterative ideas on them or are not fully negotiated or explained as being merely part-models of reality (Harrison & Treagust, 2006). It is here that drama, especially in the form of acted-out simulations, offers more plausible and accessible alternatives for understanding abstract ideas.

Learning any subject in formal education cannot be taken for granted outside a more holistic educational experience that is related to other places of learning and the learners' socio-cultural and family environments. This is the subject of my second theme.

Science and arts capital

In any discussion of the arts and sciences and EE, it is important to touch on the notion of capital, at least as space here allows. The idea of capital is often associated with the writings of Pierre Bourdieu. Bourdieu envisaged the life chances and cultural capacities of individuals being partly determined by the likelihood that they might engage in diversity and richness of cultural activities as a result of background and informal (social-family), or institutional (education-based) experience (for example, see Bourdieu, 1986). An interesting aspect here is that most of Bourdieu's writings focus on the cultural capital associated with arts and humanities activity rather than science and technology. More recently Archer, Dawson, DeWitt, Seakins and Wong (2015) have expanded on Bourdieu's notions to embrace the processes through which individuals accumulate the language, values and attributes that constitute cultural capital in science. They see part of the problem with science education being that it fails to educate the necessary habits of mind required by science. In this I would argue for an expansion of these habits to allow for the role of creativity and aesthetics which I discuss briefly in my third theme.

Research from the ASPIRES project (Archer, DeWitt, Osborne, Dillon, Willis & Wong, 2015) suggests science capital is strongly influenced by habitus from social and familial relationships and practices, often linked to socio-economic status. I would suggest here that arts capital is easier to come by than science capital and less likely to be linked with socio-economic status. Although art appreciation has a language of history and ways of seeing and interpreting, science carries more in the way of an associated set of specialised languages, symbols and artefacts, as discussed earlier. Take for example a nine-year-old presented with two artefacts. One is a painting of a flower meadow, the other a child's microscope. For the child to interact with the painting is fairly straightforward. There may be a simple reaction of like or dislike for the art, but questions of connection to it require an emotional and aesthetic reaction and simple communication, which the child will probably possess. A basic capital for art is already present. Reactions to and engagement with the microscope, to observe the flower parts from the meadow or examine the parts of insects that inhabit it, are different and require much more. Unless the child has experienced using a microscope at school or at home, it is unlikely that she or he can engage with it or appreciate its potential for illuminating the natural world of the meadow without some guidance, probably from an adult who already has the necessary capital to do so. Now, I am not suggesting that all engagement with science requires such capital input from another, but it strikes me that, as youth develop, their lives are far more likely to include the development of cultural capital that is art-based than science-based. Take for example the popular art in places like Bristol that have evolved from graffiti-tagging, once regarded as vandalism, to significant and highly valued street art, or the poetry of hip-hop and rap artists. These are aspects of modern culture more easily accessed by large numbers of youth, far more readily than for much science. So, my question is, can these forms of cultural capital for art be used to help develop science capital and impact science learning? Appreciating the aesthetics and beauty in scientific endeavour, the subject of my final theme, may also be a way of improving science capital through art involvement.

Aesthetics and beauty in science and Environmental Education

At the beginning of his book on science and aesthetics, Alfred Tauber remarks:

> Scientific experience is laden with aesthetic content of the beautiful, which is manifest both in the particulars of presenting and experiencing the phenomenon under investigation and the broader theoretical formulation that binds the facts into unitary wholes.
>
> (Tauber, 1996, p. 1)

An aspect often missing in writing about arts and sciences is the notion of aesthetics and beauty that lies in observation, ideas and theories. This is for me what underpins Tauber's comment. In his book *Imaginative Science Education*, Yannis Hadzigeorgiou (2016) points out a number of ways in which arts, and particularly aesthetics, relate to actions in the sciences. Hadzigeorgiou notes ways in which scientists feel a sense of wholeness in the unification of their ideas which represent a sense of beauty, both for the scientist and their audiences in the outcomes of the thought processes and intellectual activity that produced these theories. It may be that at the moment of creation of new ideas and theories, the boundaries between art and science cease to exist, and aesthetics play a central role. Hadzigeorgiou extends this idea, describing the role that aesthetics might play in school pupils' and scientists' experiences:

> For a third grader, for example, beauty may lie in the colours perceived during a physical phenomenon or experiment, while for a tenth grader, beauty may lie in the perception of symmetry found in nature. For a scientist, on the other hand, beauty may be found in simplicity, in symmetry, or even in complexity. It may also be found in the interrelationships of various natural phenomena and ideas.
>
> (Hadzigeorgiou, 2016, p. 190)

Of course, one could take a view that the practice of science might disrupt the aesthetic beauty seen in nature. For example, dissecting a flower might remove the beauty of the original aesthetic encounter. Alternatively, the result of dissection could add to the beauty, amazement and excitement of the observation and open possibilities for new views of the object and world. Thus, imagination and beauty can be considered the connecting ring between arts and science.

Reflecting on my experiences helping a primary teacher use wider forms of communication about investigations, I was minded that an aesthetic in science might also include affective reactions to engaging with investigative experiences such as feelings of beauty, ugliness, intrigue, pleasure, displeasure, irritation and surprise in carrying out inquiry (Anderhag, Hamza & Wickman, 2015). So often it seems that school science investigations have to come up with successful conclusions that show something definite rather than recognising that they are often messy and inconclusive (especially those in EE where the natural, living and unpredictable world is involved). Santiago Ramon y Cajal, perhaps the greatest neuroanatomist of all time, winner of the 1906 Nobel Prize in physiology and medicine, inventor of modern colour photography and painter of great talent, recognised the altruistic and artistic element of investigation:

> The investigator should possess something of this happy combination of attributes: an artistic temperament which impels him to search for, and have the admiration of, the number, beauty, and harmony of things.
>
> (Ramon y Cajal, 1951, pp. 170–171).

Conclusion

In this chapter I establish how drawing on the arts/language arts is beneficial to science and Environmental Education. I have shown how the multimodal languages of science cause problems for science learners that arts methods might moderate, that drawing on young people's art capital might also increase their science capital and that, contrary to entrenched disciplinary beliefs, there is room for beauty and aesthetics in science. As society has evolved into the fifth (post-industrial, information-based) age, there are two worrying trends. First, the nature I enjoyed in my childhood, replete with birdsong, myriad butterflies and unpolluted seas and rivers in which to bathe, is probably irretrievably degraded. Secondly, as this fifth age advances, we have moved increasingly away from a holistic learner-centred education that values science and the arts equally to one dominated by a fragmented disciplinary landscape where subject boundaries, each to be assessed and managed accordingly, seem ill-suited to prepare future generations to live effectively and be at peace in and value the world.

It seems that to equip future generations for an uncertain world – where threats like climate change and mass pandemics, once regarded as existential, are increasingly experienced in the now – will require citizens better equipped by their education to think smartly and creatively. In an article with Michael Reiss (Braund & Reiss, 2019), we made the point that, while science has been a changing landscape over the last 150 years or so, in contrast, science education has not taken much advantage of new knowledge in psychology, epistemology or pedagogy – it seems stuck with trying to represent a 19th and early 20th century, largely Baconian, version of science. In the same paper with Reiss, I pointed out the trend in the last ten years for development of large-scale enterprises and innovations, at significant cost, that link science laboratories and higher education science institutes with counterparts in the arts and humanities. In Korea and other parts of the world, there have been initiatives to widen STEM education (Science, Technology, Engineering and Mathematics) to include the A of the Arts making it, STEAM. Even so, it appears to me that, as science begins to embrace more relationships with the arts in the commercial and academic worlds, education in schools has much to do to catch up.

Acknowledgements

I was fortunate to have a mother who increased my capital in the arts and sciences through the nature walks and hiking we enjoyed together and who introduced me to the joys of music, art, poetry and theatre. I must also acknowledge my wife Shân who introduced me to the ideas of drama theorists, Brecht, Brook, Artaud, Ibsen and others. Thanks to Colin Crebbin at Bretton Hall College for exposing me to the work of Van Matre and Cornell. I continue to sit still in the places I love, experiencing the joys and wonder of nature.

References

Anderhag, P., Hamza, K. M., & Wickman, P. O. (2015). What can a teacher do to support students' interest in science? A study of the constitution of taste in as science classroom. *Research in Science Education, 45*(5), 749–784. https://doi.org/10.1007/s11165-014-9448-4

Arbuthnott, K. D., & Sutter, G. C. (2019). Songwriting for nature: Increasing nature connection and well-being through musical creativity. *Environmental Education Research, 25*(9), 1300–1318. http://dx.doi.org/10.1080/13504622.2019.1608425

Archer, L., Dawson, E., DeWitt, J., Seakins, A., & Wong, B. (2015). '"Science capital": A conceptual, methodological, and empirical argument for extending Bourdieusian notions of capital beyond the arts. *Journal of Research in Science Teaching, 52*(7), 922–948. https://doi.org/10.1002/tea.21227

Archer, L., DeWitt, J., Osborne, J., Dillon, J., Willis, B., & Wong, B. (2013). *ASPIRES: Young people's science and career aspirations age 10–14.* London: ESRC / King's College London / TISME.

Aubusson, P. J., Harrison, A. G., & Ritchie, S. M. (Eds.). (2006). *Metaphor and analogy in science education.* Dordrecht, NL: Springer.

Bereiter, C. (1994). Implications of postmodernism for science, or, science as progressive discourse. *Educational Psychologist, 29*(1), 3–12. http://dx.doi.org/10.1207/s15326985ep2901_1

Bleicher, R., Tobin, K., & McRobbie, C. D. (2003). Opportunities to talk science in high school chemistry classrooms. *Research in Science Education, 33*, 319–339. http://dx.doi.org/10.1023/A:1025480311414

Bourdieu, P. (1986). 'Forms of capital'. In J. Richardson (Ed.), *Handbook of theory and research for the sociology of education* (pp. 241–258). New York, NY: Greenwood.

Braund, M. (1999a). Electric drama to improve understanding in science. *School Science Review, 81*(294), 35–42.

Braund. M. (1999b). Using drama to improve student teachers' understanding in the physical science. In R. Feasey (Ed.), *Publication of the 4th summer conference for teacher education in primary science: The challenge of change* (pp. 13–28). Durham: School of Education, University of Durham.

Braund, M., & Reiss, M. R. (2019). The 'great divide': How the arts contribute to science and science education. *Canadian Journal of Science, Mathematics and Technology Education, 19*, 219–236. http://dx.doi.org/10.1007/s42330-019-00057-7

Bruni, C. M., Winter, P. L., Schultz, P. W., Omoto, A. M., & Tabanico, J. J. (2017). Getting to know nature: Evaluating the effects of the 'Get to Know Program' on children's connectedness with nature. *Environmental Education Research, 23*(1), 43–62. https://doi.org/10.1080/13504622.2015.1074659

Clarke, I. (2014). How to manage a revolution: Isaac Newton in the early twentieth century. *Notes & Records, 68*, 323–337. https://doi.org/10.1098/rsnr.2014.0030

Colucci-Gray, L., Burnard, P., Cooke, C., Davies, R., Gray, D., & Trowsdale, J. (2017). Reviewing the potential and challenges of developing STEAM education through creative pedagogies for 21st century learning: How can school curricula be broadened towards a more responsive, dynamic, and inclusive form of education? *BERA Research Commission: British Educational Research Association.* Retrieved October 7, 2023, from www.steamresearch.wordpress.com

Cornell, J. (1979). *Sharing nature with children. The classic parents' and teachers' nature awareness guidebook*. Nevada City, CA: DAWN Publications.

Cornell, J. (1989). *Sharing the joy of nature. Nature activities for all ages*. Nevada City, CA: DAWN Publications.

Dhanapal, S., Kanapathy, R., & Mastan, J. (2014). A study to understand the role of visual arts in the teaching and learning of science. *Asia-Pacific Forum on Science Learning and Teaching, 15*, 2. Retrieved October 9, 2023, from https://eduhk.hk/apfslt/download/v15_issue2_files/dhanapal.pdf

Gangwer, T. (2009). *Visual impact, visual teaching: Using images to strengthen learning*. London: Sage.

Hadzigeorgiou, Y. (2016). *Imaginative science education. The central role of imagination in science education*. Switzerland: Springer International/Springer Nature.

Halpine, S. M. (2004). Introducing molecular visualization to primary schools in California: The STart! teaching Science Through Art program. *Journal of Chemical Education, 81*(10), 1431–1436. http://dx.doi.org/10.1021/ed081p1431

Harrison, A., & Treagust, D. (2006). Teaching and learning with analogies. In P. Aubusson, A. Harrison, & S. Ritchie (Eds.), *Metaphor and analogy in science education* (pp. 11–24). Dordrecht, NL: Springer.

Heidegger, M., & Grene, M. (1976). The age of the world view. *Boundary, 2*, 341–355.

Henriksen, D. (2014). Full STEAM ahead: Creativity in excellent STEM teaching practices. *The STEAM Journal, 1*(2). Retrieved October 9, 2023, from https://scholarship claremont.edu/steam/vol1/iss2/15

Ison, M., & Bramwell-Lalor, S. (2023). The arts in environmental education: connecting learners with their talents and nature. *Environmental Education Research, 29*(7), 964–979. https://doi.org/10.1080/13504622.2023.2205062

Kuhn, T. S. (1970). *The structure of scientific revolutions*. Chicago, IL: Chicago University Press.

Mantere, M.-H. (1992). Ecology, environmental education and art teaching. Power of images. In L. Piironen (Ed.), *Power of images* (pp. 17–26). Helsinki: INSEA Research Congress, Association of Art Teachers.

Mantere, M.-H. (1998). Art and the environment - An art-based approach to environmental education. In L. Rubinstein Reich (Ed.), *Rapporter om utbildning, 3* (pp. 30–35). Malmö: Sweden.

Mills, D. N. (2013). *A qualitative study: Integrating art and science in the environment*. (Doctoral Dissertation, Wayne State University). Digital Commons, Wayne State, MI. Retrieved October 9, 2023, from https://digitalcommons.wayne.edu

National Curriculum Council. (1990). *Curriculum guidance 7: environmental education*. NCC. York: Pindar Graphics.

Palmer, J. (1998). *Environmental education in the 21st century: Theory, practice, progress and promise*. London: Routledge.

Ramon y Cajal, S. (1951). *Precepts and counsels on scientific investigation: Stimulants of the spirit* (J. M. Sanchez-Perez, Trans.). Mountain View, CA: Pacific Press Publishing Association.

Rauhala, O. (2003). *Nature, science and art*. Helsinki: Otava Publishing Company.

Read, H. (1946). *Annals of innocence and experience*. London: Faber & Faber.

Root-Bernstein, R., & Root-Bernstein, M. (2013). The art and craft of science. *Educational Leadership, 70*(5), 16–21. https://www.researchgate.net/publication/288657496

Ruppert, S. (2006). Critical evidence: How the arts benefit student achievement. *The National Assembly of State Arts Agencies.* Retrieved October 14, 2023, from https://nasaa-arts.org/product/critical-evidence-arts-benefit-student-achievement/

Russell, C. L. (1999). Problematizing nature experience in environmental education: The interrelationship of experience and story. *Journal of Experiential Education, 22*(3), 123–137. https://doi.org/10.1177/105382599902200304

Scott, P., Mortimer, E., & Ametller, J. (2011). Pedagogical link-making: A fundamental aspect of teaching and learning scientific conceptual knowledge. *Studies in Science Education, 47*(1), 3–36. https://doi.org/10.1080/03057267.2011.549619

Seeman, N. (2014). Art as a stimulus for structural DNA nanotechnology. *Leonardo, 47*(2), 142–49. https://doi.org/10.1162/LEON_a_00732

Shapiro, J., Rucker, L., & Beck, J. (2006). Training the clinical eye and mind: Using the arts to develop medical students' observational and pattern recognition skills. *Medical Education, 40,* 263–268. https://doi.org/10.1111/j.1365-2929.2006.02389.x

Snow, C. P. (1959). *The Rede lecture 1959.* Cambridge: Cambridge University Press.

Stafford, S. (2015). 'Pester them with letters': Using Darwin's correspondence in the classroom. In C. J. Boulter, M. J. Reiss, & D. L. Sanders (Eds.), *Darwin-inspired learning* (pp. 35–44). Rotterdam: Sense Publishers.

Tauber, A. I. (Ed.). (1996). *The elusive synthesis: Aesthetics and science: Aesthetics and science: Vol. 182.* (Boston Studies in the Philosophy and History of Science, 182). Berlin: Springer Science & Business Media.

Tytler, R., Prain, V., & Hannigan, S. (2020). Expanding the languages of science and how they are learnt. *Research in Science Education*, (July 21), 1–14. https://link.springer.com/article/10.1007/s11165-020-09952-8

Van Boeckel, J. (2013). *At the heart of art and earth: An exploration of practices in arts-based environmental education.* (Doctoral dissertation, Aalto University). Aalto Communities and Collections. Retrieved October 13, 2023, from https://core.ac.uk/download/pdf/301130226.pdf

Van Matre, S. (1999). *Earth education: A new beginning* (4th ed.). Greenville, W.VA.: Institute for Earth Education.

Wellington, J., & Osborne, J. (2001). *Language and literacy in science education.* Buckingham: Open University Press.

Wilson, E. O. (1998). *Consilience: The unity of knowledge.* New York, NY: Alfred K. Knopf.

Wittgenstein, L. (2010). *Philosophical investigations.* Hoboken, NJ: John Wiley & Sons.

4 Imagining the environment in a different, more hopeful language

(Foreign) language teaching with environmental issues and sustainability

Uwe Küchler

Introduction

Asking about the contributions of language arts to the environmental debate inadvertently leads into a discussion about the aims and principles in the fields of both environmental education and additional (foreign) language teaching: How can (foreign) language learning play a role in environmental education and education for sustainability? But there is also the question of how learning about environmental issues and sustainability can enhance and enrich language teaching, so that ecology is not just another topic added to the wide range of subjects in the language classroom. In this chapter I will therefore discuss recent publications and conceptualise the role of environmental issues and sustainability in the context of additional (foreign) language pedagogy and the exemplary discourse on English as a foreign language in Germany.

The issue of foreign (and additional) languages

While the majority of the chapters in this book refer to English as the primary or native language, they explore the role of environmental thought for contexts in which the native language is learned or taught. In native language acquisition, the learner's perception of the world is primed by and very similar to that of their caregivers and community. The "mother tongue" is a tool for communicating, perceiving and thinking. Learners are usually surrounded by that language, and even when they learn other languages or live in multilingual contexts, their native language remains a fallback option for patterns and behaviours (Saville-Troike, 2006; Thornbury, 2006; Boroditsky, 2012).

Generally speaking, any language that a speaker learns in addition to his or her first language can be considered an additional language. The terms "second language" and "foreign language" are not intended to imply a hierarchy, but they are particularly useful in characterising the social context in which learning and teaching take place. A second language is learned and experienced in the context in which it is spoken. Learners are surrounded by the

DOI: 10.4324/9781032634524-6

language they learn and need it to navigate social or professional situations in their everyday lives. But the language may not be the one spoken at home. The foreign language, on the other hand, is a practical addition to the learner's professional or social lives (Saville-Troike, 2006; Richards & Schmidt, 2002). However, it is usually not needed (in the strict sense) for personal, social or professional life. Learners use their foreign language, for example, for travel, international exchange, professional deployment or study abroad. Foreign language teaching and learning scenarios often imply that classroom activities occur in contexts that are disconnected or physically distant from the language being learned (and its community of speakers).

Particularly when it comes to environmental education, this differentiation is not trivial: while first language teachers can send their learners out into social interactions and help them integrate into society by promoting authentic interactions in that language, such a stratagem is out of reach for foreign language teachers (leaving aside the many digital opportunities or international class trips). In all likelihood, the repertoire of interactive tasks and direct environmental experience used in the classroom will be different. Foreign language teachers who live with their students outside the area of the language they are learning cannot ask their learners to use the language after class. This makes direct encounters with people, places or environments (from another part of the world) a constant challenge. Teachers of English as an additional language need to find other ways to give their students a sense of the Anglophone world – to explore landscapes or the non-human environment – and yet, not be limited to mere facts and figures.

The goals of (foreign) language education

Drastic environmental changes and the urgency to act on them have deflected attention from the language arts and the genuine goals of (foreign) language education. It would be misleading to expect the (foreign) language classroom, the teachers and language learners, to provide tangible solutions to climate change or biodiversity loss. It is all the more important now to reconsider the goals and principles of foreign language education and to rethink the discipline's self-image and its methodology. Environmental issues and sustainability do not necessarily challenge language pedagogy in terms of functional communicative needs. The discipline has long insisted that it goes beyond the functional inculcation of grammar or lexicon to include knowledge, awareness and pragmatic use of language with culture, literature and media communication.

First and foremost, the language arts and any (first, second, foreign or additional) language teaching should help learners to understand and use the language in question in any literal and concrete way (here English as a foreign language, but any language teaching may have similar goals). In addition to functional means of expression and comprehension, this goal includes the awareness of language as a communicative and symbolic system, as a tool for perceiving, understanding and even changing the world. Learners are introduced to the linguistic nature and texture of human existence in the world.

They may recognise the inextricable link between how every thought requires linguistic means, but also how language has some influence on what is perceived or noticed by speakers. This is what makes the interdisciplinary principle of language awareness (and culture awareness) such an important feature of language pedagogy (James & Garret, 1992; Boroditsky, 2012; Gnutzmann, 2017).

The strong focus on intercultural communicative competence in foreign language education argues for the close interrelationship between a language and its speakers' community, hence its culture (Byram, 1997). Much more than functional skills are needed to understand communicative symbols and successfully negotiate meanings. By identifying, differentiating, changing and coordinating perspectives, learners acquire a core ability for intercultural and transcultural learning. They become familiar with decentring their own perspectives, experiences, knowledge or attitudes and learn to understand the perspectives of others (Nünning, 2000; Bredella, 2012; Freitag-Hild, 2018).

In these learning scenarios, differences and diversity of experiences or knowledge do not make other world views better or worse. The ability to change perspectives, put oneself in someone else's shoes, coordinate one's own and many other viewpoints prepares learners for cultural encounters. It gives them the ability to think someone else's thoughts, to understand other ways of thinking patterns and reasoning. The overall goal of (foreign) language education is to provide learners with the linguistic means and communicative tools to make themselves understood to speakers of other languages, avoid a breakdown in conversation, be able to master intercultural encounters and eventually understand, read critically and comment sceptically on the arguments of others.

For these reasons, the curriculum guidelines of the German Länder build their goals squarely on intercultural communicative competence. Functional language skills – speaking, writing, listening and reading and mediation – form the basis of communicative competence and its contexts of intercultural and transcultural encounters. These skills and competences are embedded in the learners' sense of language awareness – their knowledge and experience of all language domains – and their language learning awareness – their knowledge and skills in learning and research techniques to obtain information or develop knowledge and language skills. In addition, learners' skills and competences are constantly applied to and tested against their textual and medial skills and their socio-cultural orientation knowledge (KM-BW, 2016).

Any new content or approach – such as environmental issues or sustainability in language teaching – requires foreign language educators, researchers or reflective practitioners to answer the question of what a particular topic can add to language education or, in terms of cross-curricular issues, how language teaching can contribute to such an agenda. It should make clear how a topic aligns with the goals and principles of language learning and how a case can be made that the topic is unique, special or specific to (foreign) language teaching. In other words, how can environmental issues and sustainability increase a genuine interest in (foreign) language learning and help learners understand more about language, literature, culture or media communication?

In German curricula, sustainability is seen as a cross-curricular challenge that involves all school subjects and academic disciplines. Therefore, a teaching rationale has to be considered with the interdisciplinary spectrum that language education and its related disciplines offer: in addition to environmental education, also ecolinguistics, ecocriticism or the environmental humanities have to be considered. While sustainability has found its way into the curricular guidelines, the implementation in the classroom or the specific contribution of language teaching to environmental discourse and sustainability is usually not specified. Not only the relevance of environmental issues for language learning needs to be defined and further developed, but also the relevance and contributions of the learning of (foreign) languages, literatures, cultures and media communication to wider political and economic objectives and to Sustainable Development Goals (Surkamp, 2022b; Surkamp, 2022c).

Environmental education and Sustainable Development Goals

Environmental education is concerned with the systematic and thoughtful integration of issues related to nature, the environment or sustainability into educational endeavours. The goal of environmental education is to develop an appreciation and respect for the natural world, not only in students within the school system, but in citizens of all ages and occupations. In this way, it is hoped, societies can be steered towards a more sustainable, secure and peaceful future (Palmer-Cooper & Neal, 2003). While traditionally the focus has been on natural sciences, engineering or social and political education, more recently the role of the humanities, particularly the environmental humanities, and the importance of language and communication have been considered (O'Gorman, van Dooren, Münster, Adamson, Mauch, Sörlinm ... Jørgensen, 2019).

Although the core of environmental education has long been a philosophical and pedagogical issue, geopolitical developments in the second half of the 20th century – particularly the fallout from nuclear testing, the apparent exploitation and destruction of natural environments, rampant industrialisation and globalisation – injected new energy into environmental thought. The 1960s and 1970s saw the emergence of an environmental movement and an intensified focus on environmental education (Palmer-Cooper & Neal, 2003). The field has become an umbrella term for more specific educational endeavours with different characteristics and principles, but with a shared focus on nature, the environment and sustainability. Outdoor education (Crimmel, 2003), place-based education (Sobel, 2004, Thomashow, 2002) and ecopedagogy (Kahn, 2021) are examples of such branches, all with specific aims and perspectives.

One of the common impulses for environmental education is the desire not to leave environmental issues solely in the hands of politicians or engineers, scientists, lawyers or sociologists. Instead, educators empower other disciplines and, most importantly, the learners and the citizens to contribute to solving complex problems such as humanity's relationship with the rest of the planet. UNESCO has been involved in the field of environmental education and its development since 1975, when it invited experts to the first educational workshop in Belgrade (cf. Palmer-Cooper & Neal, 2003; Dorn, 2020). When 193 member states adopted *Agenda 2030 for Sustainable Development* in 2015, they designed an agreement for the future of the world and linked ecological and environmental concerns even more closely to economic and developmental policies (BMZ, 2017; cf. also Surkamp, 2022b). The scope of the field was broadened to include more than ecological and scientific knowledge. It increasingly includes critical, ethical and creative awareness of issues related to the well-being of the planet. Ultimately, educational efforts are expected to lead to engagement and activism on behalf of the planetary environment and to changes in individual and community behaviours.

From the broad goals and premises of environmental education, one discipline has come to receive more attention: Education for Sustainable Development (ESD) has strengthened the concern with environmental issues, but it has also transformed the concern by broadening the perspective to include more societal aspects. Instead of treating certain global issues in isolation, it links research, knowledge and related issues to create a sustainable future (Leicht, Heiss & Byun, 2018). The joint efforts of Education for Sustainable Development address environmental concerns together with the fight against poverty and social inequality, the struggle for social and environmental justice, the pursuit of cultural diversity and, last but not least, the insistence on political and economic viability (Leicht et al., 2018). Particular attention has been given to educational efforts when the UN declared the years between 2005 and 2014 as the United Nations Decade of Education for Sustainable Development. These years form the basis for the articulation of the 17 Sustainable Development Goals, which link the four dimensions of economic efficiency, social justice, ecological tolerability and democratic governance (Surkamp, 2022b). These Sustainable Development Goals are now proving to be very influential and are often used as reference points when linking environmental and language education (Burwitz-Melzer, Riemer & Schmelter, 2021; Surkamp, 2022a; Römhild, 2023). In an attempt to embed Education for Sustainable Development in the German school system, Germany conceptualised an *Orientational Curricular Framework for Global Development* in 2007 and revised it in line with Agenda 2030. The focus of the educational efforts is on the learner's competences to perceive, evaluate and take action (Schreiber & Siege, 2016; Surkamp, 2022b).

Environmental issues, sustainability and foreign language teaching

While first language education embraced the field of environmental issues and environmental education early on, often administering and adapting the insights of ecocriticism or ecolinguistics (cf. Matthewman, 2011; Beach, Share & Webb, 2017), foreign language education came to the party somewhat later (cf. Küchler, 2006; Bartosch, 2019, 2021; Burwitz-Melzer et al., 2021). Yet now, the field is buzzing with research activity.

In her recent publications, Carola Surkamp has compiled premises, arguments and teaching practices that bring together and connect the Sustainable Development Goals and the curricular premises of the (German) field of foreign language education (Surkamp a,b,c, 2022). With the model proposed by Surkamp, language education has at its disposal a comprehensive pedagogical rationale that is in line with the goals and tasks of (foreign) language education and that inadvertently reaffirms the importance of a strong focus on language arts, literatures and cultures. She traces the development of Education for Sustainable Development and summarises the Sustainable Development Goals in order to develop a corresponding model for foreign language teaching. The curricular rationale is closely aligned with the UNESCO publications on Education for Sustainable Development and the curriculum for Global Development in Germany (Leicht et al., 2018; Schreiber & Siege, 2016). In what follows, I trace a model for sustainability in foreign language education that Surkamp adapts from these curricular impulses. Since the original curriculum for global development (Schreiber & Siege, 2016) neglects or leaves out some of the core competencies of foreign language education, Surkamp considerably expands the model by emphasising the close interdependence between the linguistic dimension of language learning and the cultural, literary and media dimensions. This helps to argue for the specific role of foreign or additional language education and provides a unique place for the language arts (in the broadest sense) in the environmental debate (Surkamp, 2022c).

The model is based on three aggregated competences: perceiving, evaluating and taking action (Schreiber & Siege, 2016; Surkamp 2022b).

- With the ability to perceive, learners acquire orientational knowledge and hone their analytical skills, which involve efforts to find and process information, to recognise different levels of issues or global changes (Surkamp, 2022b, 2022c). How can learners acquire knowledge and awareness about language, its dimensions and its relationship to other languages? How do cultures work? How can the interrelations between language, culture and thought, but also language variety and diversity in their cultural contexts and symbolic communicative potential be explored?
- Evaluating means above all changing perspectives and developing empathy, critically reflecting on insights and finding one's own position in relation to them. In other words: how does language affect

people? Learners recognise possible impacts of language on social interactions, including the manipulative way language is used, the interplay between language use and the exercise of power or discrimination (Surkamp, 2022b, 2022c).

- Taking action requires learners to consider solidarity and shared responsibility. They learn to understand conflicts, to respond to global changes or to participate in finding solutions. A question might be: What can people do with language? Surkamp notes that this competence remains somewhat underdeveloped and vague (Surkamp, 2022b, 2022c).

A word of caution might be in order here. The phrase "taking action" alludes to a highly motivated change in individual behaviour and finding solutions in the physical or political world. This can create misleading expectations and requires a more concise conceptualisation, as Surkamp (2022c) and Alter & Wehrmann (2022) point out. It must be emphasised that language teaching is not aimed at taking political sides, at developing scientific, technological solutions or even political activism. The German school system sticks by the Consensus of Beutelsbach (bpb, 2011), an agreement on political education or political topics in school subjects. The agreement commits teachers to avoiding political indoctrination or the overwhelming of learners with political issues, to restrain their own political opinions and to offer alternative views, thus ensuring that the range of controversial issues is presented and that learners are equipped to analyse their own (and other) societies as well as their own contexts and positions. Thus, teachers cannot and must not use the classroom to promote political engagement.

Language learning may very well provide linguistic means and a solid communicative base as well as an emotional disposition to engage in society (Surkamp, 2022c). Within the perimeter of language education, "taking action" must therefore refer to acting communicatively and interculturally. Taking action here means negotiating meaning and coordinating perspectives or navigating literature and media communication (Surkamp, 2022c). It should not mean (as might be misunderstood) political, activist solutions to environmental problems. The focus is not on inculcating a particular sustainable behaviour in learners, but rather on working on knowledge and skills as well as on attitudinal dispositions, willingness and motivation so that learners can arrive at their independent evaluation of the situation and at innovative thoughts or concepts with regard to sustainability (Surkamp, 2022b). (German) Teachers must refrain from inculcating or assessing the learners' political engagement in society but instead, learners should be empowered to evaluate and comment on sustainability discourses in the environmental, economic, political or social spheres. They should be able to take their own position or weigh perspectives or arguments in the context of cultural diversity in order to grow into responsible, future-oriented change-makers (Surkamp, 2022b).

Learning and practising these skills is embedded in different circles or contexts, which include not only the global scope most often associated with, for example, climate change, but also the national level, as well as regional concerns and local particularities (Surkamp, 2022c). Exploring language to its fullest extent, not only within the functional or cognitive domains, but also with regards to the social aspects and power domain (the affective and the performative domain), helps to reflect on aspects of language and to develop language awareness and language learning awareness (Surkamp, 2022c; Freitag-Hild, 2022).

In other words, the role of language education is much broader than providing functional means to communicate about the global issue of environmental change and sustainable living. Language and thought, culture and creativity are closely intertwined and interrelated, and learning an additional language offers complementary perspectives, helping us to think differently and allowing us to imagine other worlds. Language is what constitutes thinking and world-making (Surkamp, 2022c).

Mindsets in language teaching

Discussing the genuine goals of foreign language education and reaffirming the importance of considering language, literature, culture and media as closely and consistently interrelated, has advanced the discussion on environmental issues and the Sustainable Development Goals in the language classroom tremendously. Yet, there is another aspect that needs to be brought to the attention of educators and materials developers: the issue of mindsets in environmental education and the language arts. It cuts across curricular concerns and disciplinary principles, and it is also closely linked to the choice of topics and materials for the classroom.

Although research at the intersection of sustainability and foreign language education is only now reaching critical momentum, Surkamp points out that many environmental and sustainable topics have long since found their way into the language classroom. Foreign language education has addressed environmental issues since the 1970s (Küchler, 2016), but often the topics are chosen for their scientific merits rather than for suitability and relevance to language teaching (Surkamp, 2022c). In this way, nature, environment and sustainability enter textbooks, materials or classrooms with a one-sided focus on climate change, biodiversity loss, global species extinction or catastrophic events. All of these issues are very pressing and the urgency is undeniable. However, they are not in any obvious way consistent with the principles and objectives, content or methodology of language education. Relying on the facts and figures of environmental change not only reiterates the contents already dealt with in other school subjects, such as Biology, Geography, social studies and their bilingual branches. Arguably, the dominance of scientific knowledge in language teaching leads further away from a substantive engagement with language, culture or media communication.

The prospects for environmental situations are regularly marred by the many features of the *rebound effect:* any mitigation efforts or success stories are accompanied by greater economic cravings and ever-increasing use of mobility and tools, thus consuming more and more resources and energy. At the same time, the sense of (environmental) achievement and education is further thwarted by the sluggishness of social change and the time lag in development and widespread dissemination of new, creative, alternative mindsets. After some 50 years of environmental movements and as many decades of educational efforts in this field, it must be humbly noted that not only environmental mitigation projects but also public perceptions and awareness do not seem to be changing fast enough or far enough, nor do they seem to be triggering the necessary and required changes in behaviour on a broad social or global scale.

Such a sobering outlook should not, however, serve as a defeatist excuse. This makes it even more intractable and unrealistically difficult to expect young learners to make those necessary changes. Is it a viable solution to simply increase the pressure (on the classroom)? Environmental crises are heralded in many media outlets – only bad news is good news – and even academic articles often invoke a crisis. Of course, the debate on environmental education and the language arts has no interest in glossing over the many problems that humanity has inflicted on the environment as much as on itself. However, highly problematic environmental aspects are rhetorically foregrounded, leaving little room for any shades of optimism in the classroom (Kelsey, 2020). As if that were not bad enough, school texts and teaching materials often also pick catastrophes and their looming threats as their contribution to environmental debates. While the rhetoric of shocking people into awareness may be useful for admonishing politicians or for increasing political pressure, this mindset may not be half as useful for teachers, educators or materials developers. How are learners supposed to grow and be sustained in an atmosphere of gloom or defeat? How can educators, researchers or material developers put all the world's burden on the shoulders of young learners? This may illustrate the need for a more productive mindset and attitude on the part of educators.

When teaching about environmental issues, especially in the language arts, care must be taken not to downplay the impact of despairing mindsets, especially when considering the conflicted human relationship with the environment, but also the inner conflicts and strong emotions triggered by permanent media coverage of crises, catastrophes and emergencies. Environmental psychology ascertains that the constant confrontation with the catchwords of environmental disasters makes people feel the urge to mourn or to flee (Kretz, 2017). The enormity of climate change scenarios, for example, not to mention the multiple environmental, socio-political and cultural disruptions that will result, cannot be approached with facts, figures or rationality alone. As US psychologist Lisa Kretz warns: "The emotional impact of such losses, coupled with the alarming lack of action to remedy the situation, can lead directly to despair and avoidance if there is an absence of community support and action"

(2017, p. 259). Despair is the loss of the belief that people will change their behaviours or social structures in any committed manner. It is therefore an expression of utter hopelessness: people may feel overwhelmed by the seeming futility of all efforts to change and may therefore give up on social and environmental engagement. The loss of faith in the ability or power to make a difference can ultimately lead to the abandonment of action altogether (Kretz, 2019).

This psychological argument is supported by the classroom experiments described by US American science educator and activist Elin Kelsey (2020): when students are asked about environmental issues, they tend to think of many more catastrophic, disastrous examples than positive, hopeful images or restoration success stories. The moralistic, hopeless, often alarmist language of the media circuit – and by extension of authentic text extracts in language teaching materials – translates almost seamlessly into the despair of the general public (Kelsey, 2021). Many of the detrimental consequences of the negative and sensationalist language use and text selection are relevant for pedagogy. Even under safe circumstances, Kelsey argues, learners appear to be strongly influenced by pervasive media messages of environmental degradation. Kelsey (2020) has described the possible impact of such material upon students. They may surrender their emotions or lose sensitivity to the environment. They may even express anxieties, anger, frustration or desperation in the face of environmental problems. Such prolonged exposure to insurmountable challenges, Kelsey continues, can trigger protective mechanisms and even lead learners to deny environmental problems, thus leading to complacency or withdrawal.

The motif of catastrophe is repeated and reiterated ad nauseam in the various narratives of popular culture (Bulfin, 2017), illustrating that popular disaster stories of all shapes and sizes also reinforce people's concerns and, ultimately, their sense of apathy and fear that no change is possible. This gives the impression that human behaviour cannot be improved sufficiently to solve or mitigate environmental impacts. Narrative motifs are a strong influence on people's thoughts and actions:

> Unbeknownst to us, linguistic processes meddle in and subconsciously influence our thinking from the very basics of perception to the loftiest abstract notions and the most major life decisions. Language is central to our experience of being human and the languages we speak profoundly shape the way we think, the way we see the world, and the way we live our lives.
>
> (Boroditsky, 2012, p. 630)

The preoccupation with dystopian narratives and the gruelling effectiveness of despair is not only pointed out by psycholinguists and ecolinguists, but narratives and new media also show how powerfully expectations and belief systems shape how people think and live their lives (Bulfin, 2017; Stibbe, 2015; Kelsey, 2020, 2021). Psychological and emotional defences seem to operate in such a way that, in response to relentless exposure to impossible challenges, people are driven to deny (environmental) problems entirely (Kelsey, 2020).

For the success of environmental issues and sustainability in language education, for the motivation and enthusiasm to adapt and adopt the Sustainable Development Goals also in the language classroom, it seems to be of utmost importance to clarify, develop and nurture a mindset of hope. The importance of positive emotions has been discussed for some time in environmental psychology (Kretz, 2012, 2017, 2019) and is now also gaining ground in pedagogy and language education (Römhild, 2023). Last but not least, popular products for the general public, such as *The Jane Goodall Hopecast* (Goodcall, 2020–2022), are an attempt to take emotions such as hope or empowerment seriously. It can nourish a sense of self-efficacy and, thereby, promote individual agency as well as communal, societal change. Nancy Snow defines hope as a complex, multifaceted and interdisciplinary concept with a number of intricately intertwined factors. A hopeful mindset can be described as an attitude or a general habitual tendency, or in Snow's words, as "a dynamic orientation toward the future, characterised by the general expectancy of positive outcomes and openness toward future possibilities" (Snow, 2013, p. 153). Hopeful people adapt to and proactively respond to future opportunities as well as challenges. They maintain a forward-looking mindset and are flexible to explore new thoughts, new technologies or new approaches to making a difference. They remain therefore optimistic about what lies ahead (Snow, 2013, p. 154).

Hope relies on resourcefulness or ingenuity. As Kretz explains: "Without the capacity to imagine something other than what is currently the case, hope would have no direction" (2019, p. 161). The idea of a more optimistic outlook on the world depends as much on imagination as it does on agency. Because hope is closely related to mental and emotional states, it can develop a remarkable energy for inspiration and motivation. It thus becomes a prerequisite for a good life and for an orientation towards future possibilities. It is often driven by problematic and unsatisfactory situations. Hopefulness therefore expresses the desire for change and the belief that individuals can contribute to that change through their actions. In this context, Kretz argues that hope correlates not only with problem-solving skills but also with self-esteem. Hopeful people appear to have more control over their lives and have higher expectations of positive outcomes (Kretz, 2019).

By distancing themselves from pessimistic mindsets and by avoiding helplessness, educators could more effectively empower learners to consider their own contributions to a positive world. In her book *Hope Matters*, Kelsey showcases a number of scientific examples, initiatives and achievements that have made a meaningful difference. It would be up to language educators and humanities scholars to find and develop a comparable collection that would be suitable for working with the language arts on more hopeful environmental issues and that would also be truly motivating for the learning of (additional) languages, literatures, cultures and media.

The mindset of hope has a direct impact on the choice of content, texts and themes for the language classroom, as well as on the tasks and teaching techniques available. Educators and teachers have a crucial role to play in instilling optimism and encouraging students to embrace the future and build their lives and careers on pillars of hope, participation and agency, rather than succumbing to denial and despair (cf. Kelsey, 2020). Extending this notion to the area of sustainability in (foreign) language education, would it not be possible to envision a transformative approach? Such a transformation, however, would require a careful choice of language and storytelling that could foster a sense of environmental responsibility and inspire positive agency, and most importantly, it could direct learners towards language awareness and (trans) cultural encounters.

Examples to illustrate a more hopeful approach

Striking a balance between a hopeful and optimistic outlook and the recognition of climate change and its many consequences for the everyday lives of each individual is a bit of a challenge. One easy way for teachers to get started is to combine environmental education with a focus on language, literature, culture and communication tools. As an instantiation of such an approach, a colleague and I have put together a collection of exercises and tasks for advanced learners of English as a foreign language (Küchler & Preiß, 2020a). The workbook is complemented by a teacher's handbook that provides more detailed recommendations for teaching and task design (Küchler & Preiß, 2020b). The scope of this chapter allows for only two examples thereof.

A good starting point for classroom activities is the broader theme of language, perception and environmental awareness: How does language relate to thought, and how does it prime certain perceptions of nature and the environment? The goal of such activities would be to sharpen students' perceptive and observational skills about the world around them, the plants and animals, perhaps the seasons or any other natural cycles. They observe closely; they engage in a close, visual reading of the environment and then explore the linguistic means they have to describe what they observe. Through the study of language and perception, learners inadvertently gain a better understanding of the power of language and can extend this to other means of communication, such as visualisations, linguistic imagery and metaphorical, poetic use of language – which even occurs in environmental discourse in society or the natural sciences (for example: greenhouse effect, hockey stick graph, population bottleneck or spaceship earth).

By working with learners on popular yet still somewhat "unusual" text types or formats, such as pictures, cartoons, graphic novels or songs, the learners can experience a fascination with the linguistic framing of the world and can become active in analysing or in taking action in environmental communication. A playful approach to nature counterbalances the heavy weight that environmental issues impose on public discourse. In later instances, learners can focus not only on changing perspectives and the cultural embeddedness of perception, but also on the format and features of the discourse (cf. also Küchler, 2017).

The first example works with language that describes the natural world (cf. Küchler & Preiß, 2020a). It uses the fascinating collection of words and poems by British nature writer Robert Macfarlane in *The Lost Words: A Spell Book* (Macfarlane & Morris, 2017). Through such a poetic dictionary of words describing the natural environment, learners engage with the power of language to name, to define or to categorise the world around them. Students are introduced to selected natural words and understand the animals or plants they describe. They work with the beautiful illustrations by Jackie Morris and engage with the short acrostic Macfarlane offers for each of those words. Eventually, they are introduced to the entire "spell book" and understand that for every letter of the alphabet there is a word, an illustration and an acrostic poem, and that each of these engenders words that the English language may lose. Students comment on and discuss some features used in the acrostics that inadvertently relate forms and attributes that illustrate, mimic and recreate sounds or characteristics, such as the bickering of a magpie or the unfurling of a fern leaf (Macfarlane & Morris, 2017; Küchler & Preiß, 2020a, 2020b). These features are taken up and used in the poetic representation of this enchanted natural world, giving students the opportunity to notice features of language or the environment that they might otherwise have overlooked. Students then engage with, recite, reflect on and discuss the form of the acrostic.

To further enhance the enjoyment of these poetic forms, to go beyond the visual and make greater use of other senses, the learners play with the sound structure of the environmental word and with their own voices. They read out the poems aloud and use their voices to express different moods. What does the unfolding of a fern leaf sound like? What would make their voices resemble the bickering of a magpie? Can students find recitations that reinforce, underscore or perhaps counter, contradict or even satirise the poem? Learners discuss these sound effects and learn about the power of the human voice on the perception of the poem. These activities illuminate the domains of language and the role of language in society; they reinforce such understandings when taught in intercultural comparison, with reference to different cultural backgrounds or poetic genres. To differentiate, learners also explore other senses or go back to environmental discourse in order to see how perception, noticing, the use of senses or voices are brought to bear (or are perhaps neglected, sidelined and ignored) in popular media outlets.

The main aspect of such a task is the exploration of language and the enjoyment of linguistic possibilities. As well as noticing and becoming aware of language, students also learn about the role of language in society. It seems that it is precisely these features, often from the sensory repertoire of exploring nature, that are overlooked or neglected in environmental discourse, which does not seem to be designed for noticing (linguistic) details. While the first example pertains to the language functions of naming, categorising or mimicking the natural word, a second example works with metaphorical language. It plays with the enumeration column of the US magazine *Orion* and its often surprising environmental facts or curious juxtapositions.

The activity is based on the text "10 Things Technology Borrowed from Nature" by Sue Thomas (Thomas, 2015; Küchler & Preiß, 2020a). Students learn how technological or more abstract language can be made more understandable by drawing on physical experience or natural vocabulary, and how borrowing from embodied experience or ecology offers a wide expressive repertoire.

Students develop their vocabulary by reading about a number of words or images of nature or metaphors that have been transferred to contemporary technological contexts (cf. tree, bug, virus, cloud and more). Learners then try to describe these words and their meanings, exploring semantic features that lend themselves to equating natural with technological meanings. How do these words relate to technology? What aspects are used to transfer properties from one context to another? Students then thought-shower further words they know that may also be derived from natural contexts, but are now used in a more abstract form, in a technological sense (for example: words not used in the enumeration are (computer) mouse, tweet or stream).

By focusing on their perception, on the language at hand, on imagery and semantic transmission in language, learners not only gain language awareness. They also acquire the tools to understand how important the linguistic framing is for understanding and interpretation, how crucial such knowledge can be to help make sense of linguistic framing in any communicative contexts and how relevant such exploration is to seeing through manipulation or obfuscation that are also often used in environmental discourse.

Conclusion

In this chapter I have tried to show that language education in general, and foreign language education in particular, has a crucial role to play and a relevant voice to add to the environmental debate. (Foreign) Language teaching helps learners to develop the knowledge, skills and values they need to become environmentally responsible citizens. Language pedagogy is in the process of moving beyond the mere inclusion of environmental issues and has developed a model that reconciles the goals of environmental education and the Sustainable Development Goals with the genuine interests and principles of foreign language education. If the language classroom avoids the misconception of mixing up social activism with taking communicative or cultural action and rather focuses on communicative and (inter)cultural agency, the teaching and learning of (additional) languages, cultures, literatures and media communication can fulfil its potential as a crucial asset not only for learners' lives, but also for the wider environmental movement. Another aspect to consider is the avoidance of despair and the empowering, future-oriented energy that lies in the development of more optimistic, hopeful content and teaching approaches.

References

Alter, G., & Wehrmann, J. (2022). Auf dem Weg zu einer Kulturdidaktik des Globalen Lernens: Konzeptionelle Überlegungen zu fachexternen Impulsen und fremdsprachdidaktischen Entwicklungen. https://doi.org/10.1007/978-3-662-63782-1_6

Bartosch, R. (2019). *Literature, pedagogy, and climate change: Text models for a transcultural ecology*. Cham: Palgrave Macmillan.

Bartosch, R. (Ed.). (2021). *Cultivating sustainability in language and literature pedagogy: Steps to an educational ecology*. London and New York: Routledge.

Beach, R., Share, J., & Webb, A. (2017). *Teaching climate change to adolescents. Reading, writing, and making a difference*. London and New York, NY: Routledge.

BMZ, Bundesministerium für wirtschaftliche Zusammenarbeit und Entwicklung. (2017). *Der Zukunftsvertrag für die Welt: Die Agenda 2030 für nachhaltige Entwicklung*. https://www.bmz.de/resource/blob/23366/d52688f07df7a2c9aa78a3970295f5f5/materialie270-zukunftsvertrag-data.pdf

Boroditsky, L. (2012). How the languages we speak shape the way we think: The FAQs. In M. Spivey, K. McRae, & M. Joanisse (Eds.), *The Cambridge handbook of psycholinguistics* (pp. 615–632). Cambridge: Cambridge University Press.

bpb (Bundeszentrale für politische Bildung). (2011). Beutelsbacher Konsens. bpb.de. https://www.bpb.de/die-bpb/ueber-uns/auftrag/51310/beutelsbacher-konsens/

Bredella, L. (2012). *Narratives und interkulturelles Verstehen: Zur Entwicklung von Empathie-, Urteils- und Kooperationsfähigkeit*. Tübingen: Narr Verlag.

Byram, M. (1997). *Teaching and assessing intercultural communicative competence*. Bristol: Multilingual Matters.

Bulfin, A. (2017, September). Popular culture and the "New Human Condition": Catastrophe narratives and climate change. *Global and Planetary Change, 156*, 140–146. Retrieved April 13, 2023, from sciencedirect.com/science/article/pii/S0921818116303307

Burwitz-Melzer, E., Riemer, C., & Schmelter, L. (2021). *Entwicklung von Nachhaltigkeit beim Lehren und Lernen von Fremd- und Zweitsprachen*. Tübingen: Narr Francke Attempto Verlag.

Crimmel, H. (2003). *Teaching in the field: Working with students in the outdoor classroom*. Salt Lake City, UT: University of Utah Press.

Dorn, C. (2020). "A New Global Ethic": A history of the United Nations international environmental education program, 1975–1995. *Foro de Educación, 18*(2), 83–108. https://doi.org/10.14516/fde.808

Freitag-Hild, B. (2018). Teaching culture—Intercultural competence, transcultural learning, global education. In C. Surkamp & B. Viebrock (Eds.), *Teaching English as a foreign language: An introduction* (pp. 159–175). Stuttgart: J.B. Metzler.

Freitag-Hild, B. (2022). Kulturelles Lernen und Bildung für nachhaltige Entwicklung im Fremdsprachenunterricht. In C. Surkamp (Ed.), *Bildung für nachhaltige Entwicklung im Englischunterricht: Grundlagen und Unterrichtsbeispiele* (pp. 60–77). Stuttgart: Klett Kallmeyer.

Gnutzmann, C. (2017). Bewusstmachung / Bewusstheit. In *Metzler Lexikon Fremdsprachendidaktik: Ansätze—Methoden—Grundbegriffe* (pp. 20–21). Stuttgart: J. B. Metzler.

Goodall, J. (2020–2022). Janes Goodall's Hopecast. Retrieved November 18, 2023, from https://janegoodall.ca/hope-inspiration/hopecast/

James, C., & Garrett, P. (1992). *Language awareness in the classroom.* London: Longman.

Kahn, R. (2021). Ecopedagogy: An introduction. In *Critical pedagogy, ecoliteracy, and planetary crisis: The ecopedagogy movement* (pp. 1–33). Berkeley, CA: Counterpoint.

Kelsey, E. (2020). *Hope matters: Why changing the way we think is critical to solving the environmental crisis.* Vancouver, BC; Berkeley, CA: Greystone Books [Kindle edition].

Kelsey, E. (2021, January 18). An antidote for environmental despair. In *Hakai Magazine.* Retrieved February 11, 2021, from https://www.hakaimagazine.com/features/an-antidote-for-environmental-despair/

KM-BW, Ministerium für Kultus, Jugend und Sport Baden-Württemberg. (2016). *Bildungsplan 2016, Allgemeinbildenden Schulen, Gymnasium: Englisch als erste Fremdsprache.* http://www.bildungsplaenebw.de/site/bildungsplan/get/documents/lsbw/export-pdf/depot-pdf/ALLG/BP2016BW_ALLG_GYM_E1.pdf

Kretz, L. (2012). Climate change: Bridging the theory-action gap. *Ethics and the Environment, 17*(2), 9–27. Retrieved June 22, 2020, from https://doi.org/10.2979/ethicsenviro.17.2.9

Kretz, L. (2017). Emotional solidarity: Ecological emotional outlaws mourning environmental loss and empowering positive change. In A. Consolo & K. Landman (Eds.), *Mourning nature: Hope at the heart of ecological loss and grief* (pp. 258–291). Montreal, QC and Kingston, ON: McGill-Queens University Press.

Kretz, L. (2019). Hope, the environment and moral imagination. In R. M. Green (Ed.), *Theories of hope: Exploring alternative affective dimensions of human experience* (pp. 155–176). Lanham, MD: Lexington Books.

Küchler, U. (2006). "How Deep the Woods Are and Lightless": Seeking opportunities for environmental and intercultural teaching. In S. Mayer & G. Wilson (Eds.), *Ecodidactic perspectives on English language, literatures and cultures* (pp. 163–177). Amsterdam: Rodopi.

Küchler, U. (2016). Fremdsprachendidaktik als interdisziplinäre Wissenschaft: Ökologie, Umwelt und die Inhaltsfrage. In P. Geiss, R. Ißler, & R. Kaenders (Eds.), *Fachkulturen in der Lehrerbildung* (pp. 179–194). Bonn: University Press and Göttingen: V&R unipress.

Küchler, U. (2017). Signs, images, and narratives. In S. Siperstein, S. Hall, & S. LeMenager (Eds.), *Teaching climate change in the humanities* (pp. 153–160). Abingdon and New York: Routledge.

Küchler, U. (2021d). What's in a language: Environmental concepts and metaphors in foreign language education. In R. Bartosch (Ed.), *Cultivating sustainability in language and literature pedagogy: Steps to an educational ecology* (pp. 64–79). London and New York, NY: Routledge.

Küchler, U., & Preiß, J. (2020a). *Green matters: Planet earth-language-culture (Textheft , Schwerpunktthema Abitur Englisch—Sekundarstufe II).* Berlin: Cornelsen.

Küchler, U., & Preiß, J. (2020b). *Schwerpunktthema Abitur Englisch – Sekundarstufe II/ Green Matters - Handreichungen für den Unterricht.* Berlin: Cornelsen.

Leicht, A., Heiss, J., & Byun, W. J. (Eds.). (2018). *Issues and trends in education for sustainable development.* UNESCO. https://doi.org/10.54675/YELO2332

Macfarlane, R., & Morris, J. (2017). *The lost words: A spell book.* Toronto, ON: House of Anansi Press.

Matthewman, S. (2011). *Teaching secondary English as if the planet matters.* Abingdon and New York: Routledge.

Nünning, A. (2000). "Intermisunderstanding": Prolegomena zu einer literaturdidaktischen Theorie des Fremdverstehens: Erzählerische Vermittlung, Perspektivenwechsel und Perspektivenübernahme. In L. Bredella, F.-J. Meißner, A. Nünning, & D. Rösler (Eds.), *Wie ist Fremdverstehen lehr- und lernbar?* (pp. 84–133). Tübingen: Gunter Narr Verlag.

O'Gorman, E., van Dooren, T., Münster, U., Adamson, J., Mauch, C., Sörlin, S., ... Jørgensen, D. (2019). Teaching the environmental humanities. *Environmental Humanities*, *11*(2), 427–460. https://doi.org/10.1215/22011919-7754545

Palmer-Cooper, J. A., & Neal, P. (2003). Environmental education: International development and progress. In *The handbook of environmental education* (pp. 11–17). Taylor & Francis e-Library.

Richards, J. C., & Schmidt, R. W. (2002). *Longman dictionary of language teaching and applied linguistics*. Harlow: Longman.

Römhild, R. (2023). Learning languages of hope and advocacy—Human rights perspectives on sustainability-oriented language education. *Human Rights Education Review*, *6*(1), 9–29. https://doi.org/10.7577/hrer.5192

Saville-Troike, M. (2006). *Introducing second language acquisition*. Cambridge: Cambridge University Press.

Schreiber, J. R. & Siege, H. (Eds.). (2016). *Orientierungsrahmen fur den Lernbereich Globale Engwicklung* (In Zusammenarbeit mit Kultusministerkonferenz (KMK) und des Bundes-ministeriums für wirtschaftliche Zusammenarbeit und Entwicklung (BMZ) 2004–2015. Bonn). Berlin: Cornelsen. https://www.kmk.org/fileadmin/Dateien/veroffentlichungen_beschluesse/2015/2015_06_00-Orientierungsrahmen-Globale

Snow, N. (2013). Hope as an intellectual virtue. In M. W. Austin (Ed.), *Virtues in action: New essays in applied virtue ethics* (pp. 153–170). London: Palgrave Macmillan.

Sobel, D. (2004). *Place-based education: Connecting classrooms and communities*. Great Barrington, MA: Orion Society.

Stibbe, A. (2015). *Ecolinguistics: Language, ecology, and the stories we live by*. Abingdon and New York, NY: Routledge.

Surkamp, C. (Ed.). (2022a). *Bildung für nachhaltige Entwicklung im Englischunterricht: Grundlagen und Unterrichtsbeispiele*. Stuttgart: Kallmeyer.

Surkamp, C. (2022b). Blick zurück: Geschichte und Ziele einer Bildung für nachhaltige Entwicklung. In C. Surkamp (Ed.), *Bildung für nachhaltige Entwicklung im Englischunterricht: Grundlagen und Unterrichtsbeispiele* (pp. 7–19). Stuttgart: Klett, Kallmeyer.

Surkamp, C. (2022c). Blick nach vorn: Der Beitrag des Englischunterrichts im Kontext einer Bildung für nachhaltige Entwicklung. In C. Surkamp (Ed.), *Bildung für nachhaltige Entwicklung im Englischunterricht: Grundlagen und Unterrichtsbeispiele* (pp. 20–40). Stuttgart: Klett, Kallmeyer.

Thomas, S. (2015). Enumeration: 10 things technology borrowed from nature. In *Orion Magazine* 2015 (September–October).

Thomashow, M. (2002). A place-based perceptual ecology. In *Bringing the biosphere home: Learning to perceive global environmental change* (pp. 73–103). Cambridge, MA: MIT Press.

Thornbury, S. (2006). *An A – Z of ELT: A dictionary of terms and concepts used in English language teaching*. London: Macmillan Books for Teachers.

Part 2

Thinking about voice

5 Voicing more-than-human vulnerability

A lyric approach to fostering biodiversity loss literacy

Tanya Gautam

Introduction

The historic Kunming-Montreal Global Biodiversity Framework, adopted at the Fifteenth Meeting of the Conference of the Parties to the Convention on Biological Diversity (CBD) held in December 2022, calls for "transformative formal, non-formal and informal educational practices on biodiversity loss" that support and promote "knowledge, attitudes, values, behaviours and lifestyles that are consistent with living in harmony with nature" (CBD, 2022, p. 15). The associated Monitoring Framework also highlights the need for cultural educational practices that pay attention to the "dynamic interrelationships" between natural and socio-ecological processes, particularly in the context of "accelerated loss of biological and cultural diversity with unexpected consequences" that impact the well-being of more-than-human beings and ecosystems (UNDP and UNEP-WCMC, 2024, p. 86).

In *Climate Change Literacy*, Hoydis, Bartosch and Gurr point to the importance of reading and listening to a diversity of literary responses, stories, voices and perspectives in order to better understand the "socio-ecological complexity" of the multidimensional causes and impacts of climate change (2023, p. 38). They call on educators, and other "creators of literacy environments", to design educational tasks that pay attention to the highly complex and ambiguous nature of the impacts of climate change and invite learners to creatively reflect on their personal experiences of entanglements within their specific socio-ecological environments and to co-create, as learning groups, a "richer tapestry of complexity" (2023, p. 53). Their notion of climate change literacy considers "texts that students come up with", as a result of their engagement in creative educational practices, as "integral parts of an emerging climate change literacy" (2023, p. 59).

Meanwhile, contributions made, among others, by field philosopher, writer and environmental humanities scholar Thom van Dooren to the multidisciplinary discourse on extinction and multispecies studies underline the need for cultural practices that can help cultivate an "arts of attentiveness": of paying attention to other species – "particular ways of life" – and of meaningfully navigating one's way through the complexity of worlds unfolding at the edge of

DOI: 10.4324/9781032634524-8

extinction (van Dooren, Kirksey & Münster, 2016, p. 5). He argues that immersive practices aimed at attending and responding to the distinctive modes of knowing and being across diverse life-forms have the capacity to flesh out "thick worlds of meaning" that can "(re)craft modes of living and dying on a richly varied yet fundamentally shared world" (van Dooren, 2023, p. 3; van Dooren et al., 2016, p. 6). In doing so, he highlights the promise of literature and narratives "rich with anecdote, metaphor and figuration" (van Dooren et al., 2016, p. 8) and their potential in disclosing the complexities and ambiguities as well as continuities and connections within a more-than-human world.

In this chapter, by drawing on the work of environmental humanities scholars who call for creative educational practices on the one hand and attentiveness to more-than-human entanglements and experience on the other, I conceptualise "biodiversity loss lyricism" as a way of responding to the calls for transformative educational practices on biodiversity loss. In this light, I discuss the design process behind as well as results from a poetry writing workshop that my colleague, Katrin Klandt, and I organised as part of a cultural education initiative, *KölnerKinderUniversität*, at the University of Cologne. The initiative is directed at elementary school children between nine and twelve years old and offers lectures and workshops that introduce children to ongoing research across various faculties of the university. Our 90-minute poetry writing workshop, *What do the Dodo and T-Rex have in common?*, was guided by some of the questions currently being asked in extinction and multispecies studies. For example, "what kind of event is extinction?" (Rose, van Dooren & Chrulew, 2017, p. xii) and "what is lost when a species, an evolutionary lineage, a way of life, passes from the world? What does this loss mean within the particular multispecies community in which it occurs: a community of humans and nonhumans, of the living and the dead?" (van Dooren, 2014, p. 4). In addition to these questions, the workshop asks: How can we talk about this loss with children, including those who may not necessarily be "situated" in the multispecies communities within which the loss of many species is happening? How can creative writing, poetry in particular, help educators and children approach and understand extinction not as a singular event but as a "slow unraveling of intimately entangled ways of life" (van Dooren, 2014, p. 12)?

Specifically, the workshop and this chapter are based on a central thesis which, by speculating on more-than-human vulnerability and practices of creative writing, argues that voicing more-than-human vulnerability through the medium of poetry in general and lyric poetry in particular is a crucial step in understanding and responding to the complex and ambiguous nature of our entanglements with the more-than-human world. As Jackson argues in *The Princeton Encyclopedia of Poetry and Poetics* (2012), while the history of the lyric charts the history of poetics, lyric since its inception as a term has been used to "describe a music that could no longer be heard, an idea of poetry characterised by a lost collective experience" (Jackson, 2012, p. 826). In a time of increasing loss in life forms and their particular ways of life, this chapter asks: How can seeing lyric "not as a perfection of subjective representation" but rather as a mobile concept (Jackson, 2012, p. 832) adaptive to the unfolding entanglements and vulnerabilities of our ways of relating to one another help in creating new models of subjectivity that flesh out "thick worlds of meaning"? Vulnerability is understood here not as "a weakness, a passing indisposition, or something we can arrange to do without", but as an "underlying, ever present and abiding undercurrent of our natural state" in a co-creatively unfolding world (Whyte, 2016). In this light, I argue that voicing one's own or someone else's vulnerability, through language either spoken or written, is a way of acknowledging one's situatedness in the shifting and specific conditions of one's relationships with the diverse more-than-human worlds (Haraway, 1991). Moreover, as these worlds lurch towards the sixth mass extinction event where an exceptionally high number of species across diverse life forms are being lost in a shockingly compressed time frame (Rose et al., 2017), I explore what it means to be "situated" in the disorienting escalations of disappearing ways of life (Haraway, 1991).

Moreover, I also reflect upon and voice some of my own vulnerabilities, and what they mean to me, as a way of acknowledging my positionality: the specific "situatedness" of my life as an individual, educator and early career researcher in the field of environmental humanities. The chapter, divided into three parts, begins with this personal reflection on some of my lived and observed experiences of more-than-human vulnerabilities. The second segment focuses on the role of lyric poetry in furthering the multidisciplinary projects of extinction studies and cultural education practices on biodiversity loss. Here, I look at Craig Santos Perez's ecopoetic exploration of species extinction in his poem *The Last Safe Habitat* (2020) as its departure point to then engage with Kate Rigby's notions of negative and creaturely ecopoetics (2004, 2020) as well as Jan Zwicky's idea of lyric philosophy (2014), which lay the foundation for conceptualising what I call "biodiversity loss lyricism". The third and final segment of the chapter discusses our poetry writing workshop with elementary school children and reflects upon their poems in light of voicing more-than-human vulnerability and biodiversity loss lyricism. In conclusion, I offer reflections on the role of lyric poetry in imagining transformative formal and informal practices of cultural education that can foster what I call biodiversity loss literacy.

Voicing more-than-human vulnerability

Water

"It's not about battling against the water; it's about noticing the water and yourself in it", my swimming instructor advises. "Focusing solely on survival", he continues:

> costs a great deal of energy, but allowing your senses to perceive the water and how it moves will bring you into a relationship with it. Soon enough, you will notice how the water and your body move each other. But first, you must let go and accept that you are here, curious and willing to learn.

I do not fear water; I fear drowning. It is one of the reasons why I could not learn to swim as a child, but I am determined to learn this year, and honestly, I am getting better at it, at least every Tuesday. Though I have always been fascinated by how the water flows, shines, carries, and dries, I never immersed myself in it until recently. To learn swimming as an adult feels like learning to trust again: supported by guided learning, I am confronting my fear of drowning with a curiosity to discover and shape new experiences in and with water.

The Ferris Wheel

One of my continuing fears is to sit inside a Ferris wheel. Something about its slow circular motion, combined with height and intermittent pausing, unsettles me deeply. Instead of feeling excited the first time I sat on the Ferris wheel, I remember screaming at the top of my lungs. I remember that my siblings' promise of an awesome panoramic view did not comfort me at all. I recall commanding the operator to reverse the wheel's motion so I could get off it sooner, and he screamed back from a distance and assured me that it did not work that way. So, with hands clutching the metal bar and eyes fixed on my feet, I held on, with equally bewildered and embarrassed siblings beside me.

Almost every big city has a Ferris wheel. For me, it has become a symbol of my vulnerability: not a weakness but a way of positioning myself in the world and understanding it from this specific positionality. The more I encounter it, the more it reveals. In its slow transience, a Ferris wheel holds a collective of individuals who sign up to experience the thrill of being at the heights of vulnerability: a new vantage point. They are connected in this journey not only by their will or excitement or the rhythm of the wheel or their hearts or the force of gravity but also by their shared sense of the inability to control the forces that move or impact them; connected, temporarily, by a shared understanding of their exposure and vulnerability. Sometimes, being vulnerable in this way is an experience of both wonder and discomfort, of questioning whether something is wrong. And so every time the Ferris wheel stops unpredictably, it feels as though it has stopped working (even though it has not) and one never knows when it will move again.

The Mallard

As I write this chapter, a female mallard calls in my backyard. She's been around for the last few days and continues to fly to the roof of the building across from mine, then onto the balcony of another neighbour, and then back into my yard. She seems nervous and I initially thought that she had been injured. She refuses to eat or drink anything my neighbours and I have been offering – seeds, nuts and water. A couple of days ago, I called the local emergency number and reported an emotionally distressed and possibly injured female duck. Their immediate reaction was to send an ambulance along with an animal rescue team to my house. Upon observing the female mallard nervously flying around the backyard, the responders declared that the duck could fly just fine and did not seem to have any injuries. Instead, they reckoned she was looking for a lost mate and that it might take her a few weeks to recover from this loss. Alternatively, they explained, she could be calling to alert the males in the area about either having found a safe environment for mating or that she is nesting nearby and may, therefore, be unavailable. I will never know why or who this mallard is calling or how long she will continue to visit this particular backyard. But I know that she is a part of my world and I am part of hers. In my unsettlement and wonder about her vulnerability, I recognise also some of my own. In watching her fly in and out of spaces that I did not know a female mallard could, I was relating to her. She kept returning to voice her vulnerability or excitement or both, calling to be heard.

I did not know what to do. If it had been a lost cat or a dog, I could have posted a picture online or spread the word in my area using posters, but what does one do with a lost duck, if indeed she is lost? Who will recognise this specific female mallard and explain why she visits this particular backyard every day? The animal rescue team did not know what to do either. They asked me to call them back if the duck "continued to bother me" and that they could take her to a "shelter". I told them I was not bothered by her and that I hope she finds a way to move past this discomfort. If she indeed is looking for somebody specific – for example, the male mallard who, my neighbours tell me, was around for many weeks but is not anymore – then a forced relocation to a new address might only increase her levels of stress, right? Besides wishing I spoke her language, I am moved to think about the complex and concealed nature and future of our exchange. Similar to my experience with the unpredictable rhythm of the Ferris wheel, I do not know if and when the female mallard will return. I do not know what she is going through for certain or if it is all better at this point. Meanwhile, I realise that it is not about battling with what is being revealed in this exchange but about perceiving how the mallard moves and how at some point we indeed begin to move each other. This female mallard and her unsettled and wondrous ways of living are the waters I am still learning to notice, curiously.

These accounts are only some examples of the diverse and ambiguous ways our more-than-human lives are entangled in fragile and resilient networks. The idea behind including lyric accounts as part of an academic book chapter is also to propose a new kind of response to the calls for transformative educational practices: one that combines formal and informal modes of creative writing to not only document experiences of navigating more-than-human vulnerabilities but also to practise and contribute to the arts of noticing and cultivating attentiveness in a time of growing biodiversity loss (Tsing, 2015; van Dooren et al., 2016).

A lyric approach to cultural education on biodiversity loss

In his poem, *The Last Safe Habitat*, Craig Santos Perez (2020) voices his vulnerability as a parent in a time of growing biodiversity loss and species extinction. In his dramatically ironic lyricism, Perez declares that he does not want his daughter "to know/ that Hawai'i is the bird extinction capital/ of the world" (Perez, 2020, p. 45). He does not want her "to fear the invasive", "to see paintings/ and photographs of birds she'll never/ witness in the wild" (p. 45). He does not want her to "imagine their bones in dark museum/ drawers" or to hear "their voice recordings on the internet" (p. 45). Instead, he wants to "convince her/ that extinction is not the end", that it is "just a migration to the last safe habitat/ on earth" (p. 46).

Perez voices in his poem the private irony he practises in his protective parental vision for his child's future. The repeated voicing of denial, "I don't want to", is an attempt to move away from the dramatic irony of confronting his child with the threats of extinction. The poet foregrounds his vulnerability as a parent, precisely through the insistence of voicing it. The poet knows what the reader knows: his daughter, like all children, will learn about extinct species through photographs, paintings, and voice recordings on the internet; that extinction indeed marks the end of a species' physical existence and what sometimes survives of them are just "bones in dark museum drawers" (Perez, 2020, p. 45). In voicing his denial to confront his daughter with the fact that Hawaii *is* "the bird extinction capital of the world", Perez is also voicing his daughter's vulnerability as a protected child, the vulnerability of his parental instincts as well as that of the earth as the "last", if not the only, "safe habitat" (p. 46). Perez provokes the reader to question: Does wanting to protect your child from the reality and future of extinction render both the parent and the child vulnerable? Will the preserved and buried loss of the earth's biodiversity haunt our children and their futures? Are the paintings, photographs and voice recordings on the internet the "last safe habitat" our disappearing "winged relatives" migrate to? How, if not by memorising and reciting "the names of 77 lost species and subspecies", can children learn about extinction? (p. 45).

To begin to answer these questions, I draw upon Thom van Dooren's *Flight Ways: Life and Loss at the Edge of Extinction* (2014) where he asks:

> How might we think through the complex place of human life at this time: simultaneously, a/the central cause of these extinctions; an agent of conservation; and organisms, like any other, exposed to the precariousness of changing environments?
>
> (2014, p. 4–5)

In contrast to some recent ecocritical scholarship which suggests moving beyond lyric conceptions of the self (Griffiths, 2017; Clark, 2019), my approach to transforming educational practices on biodiversity loss dovetails with van Dooren's approach to extinction studies in that it is grounded in an "attentiveness to the diverse ways in which humans – as individuals, as communities and as a species – are implicated in the lives of disappearing others"

(2014, p. 5). Cultivating such an attentiveness in formal and informal educational settings, I argue, requires lyric work. By that I mean paying attention to and responding from one's specific and shifting situatedness in an increasingly vulnerable more-than-human world. To do lyric work in a time of growing biodiversity loss, what I call "biodiversity loss lyricism", is to give voice – literal, metaphorical, figurative or gestured – to an "ecology of experience" with an aim to foster creative and critical modes of literacies that shape formal and informal educational ecologies of resonance and attunement (Zwicky, 2014, p. 16). Germaine to this "biodiversity loss lyricism" is Rigby's notion of "creaturely ecopoetics" which is rooted in attending to human and non-human entanglements that entail:

> shared, if unevenly distributed, vulnerabilities, as well as shared, if variegated, communicative capacities; entanglements that harbour the ever-present risk of conflict and harm, as well as opportunities to co-create emergent multi-species worlds no longer constrained by the colonizing logic of human-nonhuman hyperseparation, and hence freed to explore more felicitous forms of coexistence in risky times.
>
> (2020, p. 84)

Lyricism, by offering such open and adaptive forms of exploration, is a crucial step in understanding and approaching vulnerability not as an incompatible notion of existence but as a conversational mode of being that needs both informal and formal practices of cultural and educational engagement, which often lack the means of discussing vulnerability, to cultivate capacities and ecologies of resonance on a number of socio-ecological dimensions. Like Jan Zwicky in *What is Lyric Philosophy: An Introduction*, I argue that practising (biodiversity loss) lyricism rooted in metaphor, affect, figuration and relationality is one way of discerning the shifting relations between human and non-human others which can further foster a perception of deep entanglements and vulnerabilities which while overriding the distinctness of individuals does not aim to eliminate or erase the self (Zwicky, 2014). A biodiversity loss lyricism is one that is also aware of the limits and vulnerability of language. Both Rigby and Zwicky consider language to be "a limited instrument – vast, supple, complex but limited" (Zwicky, 2014, p. 19). In pointing to the inadequacy of language – "negative ecopoetics" – to fully articulate the complex and often concealed more-than-human entanglements, Rigby calls for a plurality of voices: "for if no one can say it all, then we are all called upon to participate in our own way in the 'chorus and polyphony' of responses" (Rigby, 2004, pp. 438–439). Our poetry writing workshop *What do the Dodo and T-Rex have in common?* is an example that shows how practising biodiversity loss lyricism in educational settings is a way of generating a plurality of voices and transforming educational practices on biodiversity loss, thereby fostering a more nuanced understanding of what can be called a "biodiversity loss literacy".

What do the Dodo and T-Rex have in common?

As discussed in the sections above, our approach to species extinction was informed by an understanding that extinction is not simply an abrupt absence of the "body and genetic material" of a species (Rose et al., 2017, p. xi). It is a disruption in the "intergenerational lineage" of a species: a whole way of life that is cultivated over millions of years (Rose et al., 2017, p. 9). Extinction is a multidimensional phenomenon that is "experienced, resisted, measured, enunciated, performed, and narrated in a variety of ways" (Rose et al., 2017, p. viii). The specific nature and conditions of disappearance that these "whole ways of life" experienced often remain concealed in multicontextual layers to which we must attend and respond.

What the Dodo and the T-Rex have in common, then, are the layered mysteries of their disappearance and the complex ways in which it impacted other life forms and ecosystems. While the fifth mass extinction event wiped out, among others, all non-avian dinosaurs like the T-Rex, it also paved the way for the evolution of new life forms and ecosystems we know today (Alvarez, Álvarez, Asaro & Michel, 1980). On the other hand, the human-driven extinction of the Dodo triggered lasting impacts on Mauritius' ecosystem dynamics and biodiversity. It led to cascading declines in many of the plant and animal species on the island whose plant-animal networks continue to face high risks of co-extinction (Heinen, Florens, Baider, Hume, Kissling, Whittaker, Rahbek & Borregaard, 2023).

Our poetry writing workshop addressed some of the diverse causes, impacts and meanings of extinction, both as a historical event and as an ongoing process around the world. To engage elementary school children in a nuanced exploration of extinction, we structured our approach focusing on three themes: listening to sonic diversity; attending to evolutionary and socio-ecological connections; and crafting "worlds of meaning" through poetry.

Listening to sonic diversity

In *Sounds Wild and Broken: Sonic Marvels, Evolution's Creativity, and the Crisis of Sensory Extinction* (2022), biologist and writer David George Haskell addresses the impacts of climate change on sonic diversity worldwide. In light of increasing habitat destruction and noise pollution, Haskell argues that the disappearance of wild sounds renders the world more vulnerable and less generative and results in an "impoverished sensory world" (2022, pp. xiii–xiv). Listening to sonic diversity, he argues, not only "opens us to the wonders of communication and creativity" but also "teaches us that we live in an age of diminishment" (2022, p. xiv). Listening to the earth's sonic diversity is to engage, aesthetically, with the perceptual crisis in which we live. Practising such an engagement is crucial for building human ethics and fostering biodiversity loss literacy because:

> When the most powerful species on Earth ceases to listen to the voices of others, calamity ensues. The vitality of the world depends, in part, on whether we turn our ears back to the living Earth.
>
> (2022, p. xiv)

To address this complex crisis of sensory extinction in a creative and educational way for elementary school children, we conceptualised an immersive game called Mystery Sounds. In this game, children were introduced to audio-visual prompts featuring species categorised across IUCN's extinction risk levels, ranging from Extinct to Least Concern and the objective was for children to associate the sound with the corresponding animal. Each iteration displayed photographs of three animals on the screen, for example, a German Shepherd, a Common Pigeon and an Orangutan. We then played a 30-second audio recording of one of these animals and asked children to position themselves under the photograph of the animal they correlated the sound with. It is interesting to note here that for one of the iterations the visual prompt included the extinct Kauai O'o bird, a crow and a Pygmy blue whale. The audio we played for this round was the last recorded call of the Kauai O'o bird of Hawaii. While some children chose the Kauai O'o, most stood under the whale.

Attending to evolutionary and socio-ecological connections

Ongoing processes of biodiversity loss and extinction are unfolding rapidly in compressed time-frames on various spatial scales (Rose et al., 2017). In order to understand these rapid scales of disappearance and loss, it is important we see extinction as an "extinction of experience" (Pyle, 2012, p. 136). In order to fully understand the consequences of loss in life forms and interactions and how they get translated into experience across other life forms, it is crucial to juxtapose the various scales at which extinction takes place. This not only discloses the continuities and connections of its cause and effect across ecosystems, networks and human cultures but also points to the "nonscalability" (Tsing, 2012, p. 505) of the more-than-merely-human experience of loss and impact.

To attend to these ideas of temporal and spatial scales during our workshop, we prepared worksheets, each of which included information on two species or subspecies, for example, the Dodo, and its closest living relative, the Nicobar Pigeon. By attending to extinct species in comparison to related species that are threatened by extinction today, our goal was to offer children a sense of relationality and continuity in processes of both evolution and extinction. Along with addressing extinct species and their closest living relatives or species who continue to remain threatened by, among others, habitat destruction, poaching, overhunting, pollution, invasive species and climate change, we also thought it crucial to speak of recovering species that have seen an increase in population as a result of conservation efforts, such as the American bison.

After reading the information we provided, children were also asked to answer the following questions on their worksheet: Why did this animal go extinct? Which sounds do you imagine this animal made? What more do you know about them? By asking these questions our aim was not only to establish a clear learning goal for children participating in the workshop but also to take into account their prior knowledge regarding the animals they were engaging with.

Crafting worlds of meaning

For the creative writing segment of the workshop, we first presented the forms or techniques of poetry we chose as prompts: haiku, elevenie, acrostic, anaphora and found poetry. We chose these particular forms to cater to the age group we were working with and to give as many examples as possible that support the different ways children might approach writing. While the haiku, elevenie and acrostic have some relatively fixed elements to support structural writing, anaphora and found poetry were chosen to support a more freestyle writing approach. To begin with, we read examples I wrote especially for this workshop. Knowing I was writing poetry for young children, my aim was not to speak of a complete or specific story or idea but instead to inspire a sense of play, mystery and ambiguity. Most importantly, my goal was to make sure that children found poetry accessible and not too complicated to attempt writing, including examples that encourage participation, play and response. Moreover, to showcase different ways one could write about the same animal, the poems I wrote as introductions to the poetic forms and techniques were all about snakes:

Anaphora

Sssss, says the king cobra!
Sssss, seeks the king cobra!
Sssupper time in the jungle,
Who isss ready to join? There will
Alwayss be enough ratsssss for uss all!

Found Poetry

Shy creatures, we are!
Love to hide, we do!
Every child is unique,
Just like every snake.
Would you like to get to know me better?
Then wait here in the serpentine,
Eh ... you're going the wrong way!

Haiku

Hiking in April.
Snake skin shining in sunlight.
The snake must have moved!

Elevenie

Rattlesnakes in the desert.
Poisonous. Cunning.
Who is winning?
Future – unknown.

Acrostic

See me yet? Pay attention and listen closely.
Nice and slow I move along the forest floor. No, I
Am not a leaf in the wind but a
sil**K**y
Elegant reptile.

All poems written for or during the workshop are in German and are included here in their English translations. This too points at the vulnerability of language, of translating it at the risk of losing rhyme, structural creativity and the ambiguity of language. But for the purpose of this chapter, I focus on foregrounding the vulnerabilities voiced in children's poems and how creative writing as a participative approach is able to cultivate, from early on, habits of attending and responding to the disorienting escalations of disappearing ways of life. The poems children wrote point to the importance of engaging with the temporality and spatiality of extinction and what it means for children to put together the diverse and ambiguous ways these processes unfold for different species across the history of life and time on Earth. Noticing the diversity of response in children's poems, in particular, foregrounds the importance of creative modes of writing in educational practices that allow attending to and voicing unique and emerging ways of relating to more-than-human others in a time of growing biodiversity loss.

Evident in children's poems, for example about the Megalodon, is how differently children imagine extinct species and their ways of life and what it means to relate to them as humans. While one child talks about the Megalodon as a fine shark that could "fit an elephant ten times", the other speaks of the "gigantic bite that is frightening" and how "humans can sometimes be a feast". They imagine a "fearless" and "mega strong" Megalodon, a "maverick cruising all alone" in his dimly lit territories and how the changing climates in the seas drove it to extinction: "the seas got colder" and "the last of its kind grew older". Engaging with extinction as well as threats of extinction through the creative play of poetry illuminates children's emerging knowledge of ecological connectivities and poetic sensibilities as well as their vulnerability as children who did not have a chance to save the animals they find so fascinating. About the extinct Tasmanian Tiger, a child wrote:

> Whiskers that sense you,
> From afar, I smell you.
> His coat is richly striped.
> They saw me in forests.
> From forests grew cities.
> Of course, we too die.
> I didn't live very long.
> His time had come and
> Chances do not exist.
> Listen, I'm already dying.
> A life turns to stone.
> The tiger returns home.
> I lived alone in a zoo
> And my species was shot dead too.
> He died alone.
> I would have liked to save him.

In her poem, the child voices her desire to have had a chance to save the Tasmanian Tiger from going extinct. She creates an ambiguity of voices and vulnerabilities: her own and that of the tiger, the forests, cities as well as the vulnerability of chances followed by the alarming call to "listen, I'm already dying". She imagines a life turning to "stone" as the "tiger returns home" and points to the routines of "violent care" (van Dooren, 2014, p. 87) the tiger must have witnessed when he "died alone in the zoo". This is an example of how aware some children are of the consequences of, and urgency to act against, socio-ecological violence towards animals. Here is an alarming call to listen to the disappearing life as well as a dramatically ironic assertion that "chances do not exist": pointing to the lack of opportunity felt by a 10 year old child who wants to help avoid extinction.

Interesting to note was also the difference in the tone of urgency with which children voiced their imaginations and concerns. As seen above, poems about the Megalodon are less alarming in their tone than in the poem about the relatively recent extinction of The Tasmanian Tiger. Other poems, for example those about the endangered Oceanic Manta Ray, were less urgent and more playful in their approach while also calling readers to be fascinated by the "lovely giant manta ray". In a dreamy riddle about the Manta Ray, a child wrote:

Me, I am not.
You, I am also not.
Of course it's not yet extinct!
Are they dreaming?
Answer this question, I can not.

The ambiguity of the fourth line is more evident in the original German version of the poem. "Träumen sie den?" could be addressing the reader to ask if they ("Sie") are dreaming about the Manta Ray being extinct and reminding them that it is not. The poet could also be asking if the Manta Rays (sie) too can dream? Or is the reader dreaming *of* the Manta Ray? Finally, it could just be a misspelt "denn" in German but that's the beauty, I believe, of ambiguity and poetry written by children (and in general): it is not necessary to narrow down on a singular meaning or intention.

The distinct, more-than-human vulnerabilities voiced in these poems are examples of what I have called a "biodiversity loss lyricism". They point to habits of imagination and response and also show how children relate with animals in their sociocultural and emotional ways of being. For example, another child wrote about the "lovely creature" Manta Ray who, despite the contradictory nature of his movements and luck, is loved unconditionally by his sister:

Sometimes he would swim,
And sometimes he would jump!
Sometimes he'd catch a fish,
And sometimes it would be a miss!
Sometimes he was fast,
And sometimes he was slow!
That's just how giant manta rays are!
And his sister would say,
Even when no one would ask,
That he's a lovely giant manta ray!

In their poems about the Sea Turtle, children speak of a "beautiful", "lovely", "talented", "lonely", "very intelligent" and "almost extinct" creature that continues to be hunted for its flesh and shell, a "thick" and "suffering shell" which is "round like this globe". They implore the reader to save the turtle as "there is still enough time".

Finally, at the end of the workshop we gathered in a semi-circle to read our poems out loud as a way of voicing our response, vulnerabilities and situatedness. It was especially interesting to see that children encouraged each other to read their poems out loud and that at the end of the session when we asked children for their feedback, they expressed their joy in having found others who share their interests and observed how differently they each expressed their thoughts.

Conclusion: towards a biodiversity loss literacy

The poetry writing workshop, *What do the Dodo and the T-Rex have in common?*, is one example of the transformative educational practices that can be imagined at the nexus of environmental humanities, biodiversity loss and lyric poetry. Incorporating formal and informal modes of creative writing in education, I have argued, can contribute to emerging ecological knowledges and poetic sensibilities in a time of growing biodiversity loss. Drawing on my own vulnerabilities in connection to the poetry writing workshop and children's poems, as well as on the ongoing contributions in the broader field of environmental humanities, in this chapter, I suggest that literary and cultural responses to the "extinction of experience" should engage with extinction as socio-ecological crises of perception and cognition as well as the capacity to respond. Voicing more-than-human vulnerability through what I call "biodiversity loss lyricism" has the capacity to address specific and shifting more-than-human situatedness and to cultivate ecologies of experience and resonance in a time of disorienting escalations of disappearing ways of life: an approach that is aware of the limits of experience and language as the world lurches towards the sixth mass extinction event.

References

Alvarez, L. W., Álvarez, W., Asaro, F., & Michel, H. V. (1980). Extraterrestrial cause for the Cretaceous-Tertiary extinction. *Science, 208*(4448), 1095–1108. https://doi.org/10.1126/science.208.4448.1095

CBD. (2022). Kunming-Montreal global biodiversity framework. In *Conference of the parties to the convention on biological diversity*. https://www.cbd.int/doc/decisions/cop-15/cop-15-dec-04-en.pdf

Clark, T. (2019). *The value of ecocriticism*. Cambridge: Cambridge University Press.

Gaston, K. J., & Soga, M. (2020). Extinction of experience: The need to be more specific. *People and Nature, 2*, 575–581. https://doi.org/10.1002/pan3.10118

Griffiths, M. (2017). *The new poetics of climate change: Modernist aesthetics for a warming world*. London: Bloomsbury Academic.

Haraway, D. J. (1991). *Simians, cyborgs, and women: The reinvention of nature*. New York, NY: Routledge.

Haskell, D. G. (2022). *Sounds wild and broken: Sonic Marvels, evolution's creativity, and the crisis of sensory extinction*. London: Penguin.

Heinen, J. H., Florens, F. B. V., Baider, C., Hume, J. P., Kissling, W. D., Whittaker, R. J., Rahbek, C., & Borregaard, M. K. (2023). Novel plant–frugivore network on Mauritius is unlikely to compensate for the extinction of seed dispersers. *Nature Communications, 14*(1). https://doi.org/10.1038/s41467-023-36669-9

Hoydis, J., Bartosch, R., & Gurr, J. M. (2023). *Climate change literacy*. Cambridge: Cambridge University Press. https://doi.org/10.1017/9781009342032

Jackson, V. (2012). Lyric. In R. Greene (Ed.), *The Princeton Encyclopedia of poetry and poetics* (pp. 826–834, 4th ed.). Princeton, NJ: Princeton University Press.

Perez, C. S. (2020). The last safe habitat. *Habitat threshold*. Oakland, CA: Omnidawn.

Pyle, R. M. (2012). *The thunder tree: Lessons from an urban wildland*. Corvallis, OR: Oregon State University Press.

Rigby, K. (2004). Earth, world, text: On the (Im)possibility of ecopoiesis. *New Literary History, 35*(3), 427–442. http://www.jstor.org/stable/20057847

Rigby, K. (2020). *Reclaiming romanticism: Towards an ecopoetics of decolonization.* London: Bloomsbury Academic. http://dx.doi.org/10.5040/9781474290623

Rose, D. B., van Dooren, T., & Chrulew, M. (2017). *Extinction studies: Stories of time, death, and generations.* New York, NY: Columbia University Press.

Tsing, A. L. (2012). On nonscalability. *Common Knowledge, 18*(3), 505–524. https://doi.org/10.1215/0961754X-1630424

Tsing, A. L. (2015). *The mushroom at the end of the world: On the possibility of life in capitalist ruins.* Princeton, NJ: Princeton University Press.

UNDP and UNEP-WCMC. (2024). *Using spatial data to support the development of plans for national monitoring systems for the Kunming-Montreal global biodiversity framework.* Report prepared for the Global Biodiversity Framework Early Action Support Project, in collaboration with the UN Biodiversity Lab Partnership. https://unbiodiversitylab.org/wp-content/uploads/2024/04/EN_Guidance-Using-Spatial-Data-to-Support-GBF_29-April-2024-1.pdf

van Dooren, T. (2014). *Flight ways: Life and loss at the edge of extinction.* New York, NY: Columbia University Press.

van Dooren, T. (2023). Worlds of meaning at the edge of extinction: Conservation behaviour and the environmental humanities. *Humanities, 12*(5), 122. https://doi.org/10.3390/h12050122

van Dooren, T., Kirksey, E., & Münster, U. (2016). Multispecies studies: Cultivating arts of attentiveness. *Environmental Humanities, 8*(1), 1–23. Durham, NC: Duke University Press.

Whyte, D. (2016). Vulnerability. *On Being.* https://onbeing.org/blog/vulnerability/.

Zwicky, J. (2014). What is lyric philosophy?: An introduction. *Common Knowledge, 20*(1), 14–27. https://www.muse.jhu.edu/article/532879

6 Philosophical dialogue in ecoliteracy education

Words for change

Jelle de Schrijver and Lynda Dunlop

Introduction: language and philosophy to address ecoliteracy

It is well established that climate change is happening as a result of human activities, with unequal historical and ongoing contributions to greenhouse gas emissions as a result of unsustainable energy use, land use and lifestyles and consumption practices (IPCC, 2023). This has profound impacts on human and planetary health and well-being, with increased hazards due to extreme heat-, food-, water- and vector-borne diseases, flooding, biodiversity loss and food production. Education can play a vital role in the way societies respond to heightening climate and ecological crises in terms of raising understanding of ways to reduce greenhouse gas emissions and adapt to climate change, and in creating a public mood conducive to policies that attempt to reduce greenhouse gas emissions. Although climate education can be effective at reducing emissions (Cordero, Centeno & Todd, 2020), education can also contribute to unsustainable living due to a "lack of opportunity for learners to question their own lifestyles and the systems and structures that promote those lifestyles" (UN Economic Commission for Europe, 2012, p. 6). In this chapter, we will examine this tension between what is and what could be by examining the role of philosophical questioning and dialogue in language arts education.

Literature and language are powerful tools in (environmental) education. The philosopher Martha Nussbaum pointed out that literature helps students cultivate empathy and allows readers to engage with diverse perspectives and moral complexities (Nussbaum, 1990). In a similar vein, climate fiction, climate change theatre and ecopoetry provide opportunities to re-think societal structures and patterns and explore future sustainability in a safe space (Leavenworth & Manni, 2021; Johns-Putra, 2016). More broadly, language gives learners the tools needed to listen, speak, read and write, all of which are necessary to think about, create, collaborate and communicate about climate change, how to act most effectively in the present and imagine more desirable futures. Moreover, language provides tools to think critically, think about the words and ideas we use, consider the options we have and discuss what kind of actions we could or should take. Questioning and critique are key to instruction in the language arts, and also to developing ecoliteracy, which is

DOI: 10.4324/9781032634524-9

concerned with developing an understanding of the world and ecosystems and taking action in and with the environment (McBride, Brewer, Berkowitz & Borrie, 2021) and applying principles of organisation of ecosystems to create sustainable human societies (Orr, 1992). This brings us to the domain of philosophy; after all a focus on critical thinking and reflecting on the fundamental questions underlying our relation to the environment is key to it.

Philosophy and language are closely intertwined. Philosophical ideas are shared and challenged through language transmitted orally and in written form through generations. Language is not only a communication medium; it can also be an instrument to unravel multiple meanings and interpretations. In the dialogue *Phaedrus* written by Plato, Socrates argues that the first rule of good speaking is to know and speak the truth, contrasting dialectic speech from rhetoric, which, although it has power in public assemblies, Socrates describes as an "art of enchanting the mind" (Plato, 2013). The former is described as speech based on knowledge of the truth which can be defended when put to the test by written or spoken arguments. The ability to distinguish between truth and rhetoric is especially important in the context of climate change, where communication happens in a context of high political polarisation, superficial public understanding and growing hopelessness in the face of government inaction (Moser, 2016; Hickman, Marks, Pihkala, Clayton, Lewandowski, Mayall, Wray, Mellor & van Susteren, 2021).

In the 1990s, Fritjof Capra and David Orr introduced the concept of *ecoliteracy* as the ability to understand the natural systems that make life on earth possible and to use those principles for creating sustainable human communities (Orr, 1992). Ecoliteracy requires an understanding of the interconnectedness of life and connection and empathy with living things (Wallace, 2019). Thus, it was argued that understanding and caring for the well-being of the earth should be a curriculum aim. Drawing on an analysis of different definitions and frameworks for ecoliteracy used since the 1990s, McBride et al. (2021) define ecoliteracy as when someone is "prepared to be an effective member of sustainable society, with well-rounded abilities of head, heart, hands, and spirit, comprising an organic understanding of the world and participatory action within and with the environment" (p. 67). Yet, a focus on ecoliteracy in schools can be demanding to enact. In this chapter, we explore three challenges with ecoliteracy education that can be tackled by focusing on philosophy and language in education. The first relates to learning to grasp the meaning of words used when we think about the environment and sustainability. The second focuses on the argumentation skills necessary to understand and think about environmental and related social issues. The third entails the challenge to create a shared dialogue and a co-thinking space where students can discover and examine each other's perspectives. As we address these challenges, we explore how the classroom practice of "philosophical dialogue" can give teachers and learners the power to reflect on climate change, challenge existing systems and create new ways of being through language. We outline how this might happen within a classroom context, give examples of philosophical dialogue in environmental contexts and then discuss some of the tensions that arise when doing philosophical dialogue with a given purpose.

Challenges in the field of ecoliteracy education

The meaning of words

Can an activist change the world? Can an energy company be sustainable? Does it matter if the earth becomes inhospitable to humans? What does it mean to rewild? Answering these questions can be challenging and exciting. The answers depend on how we understand the words that are used. This brings us to the field of semantics – the study of the meaning of words. Words like "sustainability", "nature", "activism", "wilderness" and even words like "better" or "change" can have many meanings and connotations, and can change. The meanings of "atmosphere" and "climate" (both literary and literal, or figurative and technical), for example, have changed over time (Ford, 2022). Often challenges in climate communication (and education) result from differences in emphasis, meaning or interpretation of words and frames (Bromhead & Goddard, 2023) or from the ideological framing that words carry in relation to nature, culture, society and different knowledge systems (Hecht, 2017/2021). For instance, a company can call itself sustainable if it aims for a sustainable future for itself, yet this kind of sustainability is not necessarily sustainable for the planet. The meaning of words such as "natural", "organic" and "sustainable" used in discussions about environment or climate change is often disputed and imprecise. Furthermore, the scientific meaning of words can be mixed with the everyday use of words, and to clarify these issues, the teacher has a role to play to bridge scientific and everyday language (Bravo & Chion, 2017). For instance, for a biologist, the word "nature" may denote the natural living environment including humans. In everyday language "nature" or "natural" is often used as precisely the opposite of "artificial", and humans are placed on this "non-natural" end of the opposition. In this regard, unravelling the meaning of the words, and exploring the meaning of the words in a particular context, is a key competence for understanding the questions surrounding the debates on climate, environment and sustainability.

The challenge is to teach students how to uncover the meaning of words. This asks for attention towards language, on how the meaning of words can change in different contexts and towards the conceptual (philosophical) meaning of the words themselves. Philosophical concepts have been described as common, central and contestable (Splitter & Sharp, 1995); in other words, philosophical concepts such as truth, justice and knowledge are familiar to most people and are central to what it means to be human. This multiplicity of meanings and connection to the human condition means that philosophical dialogue is accessible to learners. The philosopher Mary Midgley observes that philosophy is concerned with the complex but often invisible structure underpinning knowledge, society and culture (1992). Midgley notes that philosophy, like plumbing, is needed, especially when things have gone wrong, as they have in the case of the impact of anthropogenic activity on the earth's systems such that six of the nine planetary boundaries within which humanity can thrive for generations have been transgressed (Richardson, Steffen, Lucht, Bendtsen, Cornell, Donges & ... Rockström, 2023). By focusing on philosophy and exploring the meaning of words, we can teach students to think about the key (value-loaded) questions that underlie the "plumbing" of the world we live in and help students understand and explore together how we might possibly respond to the environmental challenges we face. This requires those participating in dialogue to examine and adjust assumptions and identify hidden and different meanings.

Language, logic and argumentation

The close connection between language and thought can be misleading as language may also be the source of deception. Using rhetorical tricks to convince the audience is an age-old art that has been employed by speakers and writers throughout history. Persuasive techniques are methods used to influence and manipulate an audience's emotions, beliefs or behaviours. These can include employing powerful imagery, appealing to the audience's emotions, and employing figures of speech to enhance the persuasiveness of a message. For example, "greenwashing" is a practice that falsely represents an organisation's environmental record by, for example, making vague or relative claims ("uses 90% more recycled materials"), using imprecise words ("green", "eco-friendly"), using green or natural imagery or using unrelated information. It is important to question such claims: what is it you are being led to believe? Is a specific claim being made, and what evidence exists to support it? After all, the use of rhetoric raises questions about manipulation and truthfulness, also in the domain of climate change, and its impact on the environment. This is familiar in many practical contexts, perhaps especially in politics. The same danger exists everywhere, however, including in scholarly and scientific research. Learning to use, analyse and identify these rhetorical instruments is a challenge for both teachers and students alike.

The Greek philosopher Plato was characteristically wary of the power of words, which the Sophists relied upon. Hence his mistrust of rhetoric and his banishment of poets and artists from the ideal state he described in *The Republic*. He preferred to think instead of the naked encounter of the properly trained mind with the essence of things. One does not need to share Plato's ideas about essentialism to see how his warning echoes to this day. Philosophy's focus on logic and argumentation precisely helps analyse and (de)bunk misleading reasonings. By focusing on logic, students learn what a good argument is and how a conclusion does (not) follow from its premises. A focus on logic and argumentation provides students with an understanding of reasoning and the vocabulary to discuss the truth and reliability of claims. This gives students power to read the world, and language to call out fallacious arguments and false claims. Thus, in the context of environment and climate education, it is relevant and meaningful to help students focus on the logical structure of language to analyse how convincing or misleading arguments can be constructed.

Language and shared or "dialogic" thinking

Tackling climate change cannot be done by one individual; it depends on cooperation, deliberation and dialogue. Yet, creating a climate for dialogue can be challenging in itself. Tom Berryman and Lucie Sauvé point out that individuals know, think and act according to interiorised culturally inscribed paradigms. This includes paradigms about how the world functions, how decisions are made and how our lives should be lived. These culturally inscribed paradigms can lead people to live in echo chambers (Berryman & Sauvé, 2016), making it hard for people to see beyond the world as they know it. In this regard, education can play a key role in helping students examine their lives and explore the different ways to think and act in the world. A paradigmatic discourse, with its specific use of language and argument, can be deconstructed. In doing so, it asks for a confrontation with different kinds of thinking. The challenge is to create an atmosphere of dialogue wherein students are invited to explore broadly accepted paradigms and each other's points of view. Shared philosophical dialogue across differences has been used to understand what children think about genetic screening (Fields & Asbury, 2023), indicating the accessibility of philosophical dialogue for the consideration of controversial or contested topics, even amongst very young children.

There is a need to learn to think together about global challenges such as climate change, as only by engaging in shared dialogues can students learn how to develop a sound understanding of (the political, social and ethical aspects of) climate change. The challenge is to speak together, even with people who have different views on what we might need to make the world a better place. Indeed, in describing connection with others, Hermans and Bartels (2020) argue that there is a need for the personalisation of democracy, in other words, to make citizenship an integral part of our lives, and that dialogue requires flexible boundaries with others (where one knows their position but is open to the perspectives of others) in contrast to the rigid boundaries that are associated with social stereotyping. Philosophical dialogue helps to create greater flexibility through making connections with others through dialogue.

To help students develop a "dialogical capacity", several aspects are considered relevant including perspective-taking and having constructive disagreements with others (Hermans & Dimaggio, 2007). Again, practical philosophy may be relevant to tackle this challenge. Dialogue is characteristic of (at least part of) the philosophical tradition. Not only have some of the philosophical works been written as dialogues (for example Plato's or Hume's); philosophy and the arts (just like science) are built out of a dialogue between contemporary thinkers and bygone voices. Furthermore, philosophical practice provides tools that can be used to introduce critical thinking and dialogue in education.

By engaging in philosophical dialogues, students are enabled to explore multiple perspectives on philosophical issues. This approach can help bridge the divide between theory and practice and help students reflect on the relationship between words, feelings and actions concerning climate change and other planetary boundaries. Thus, philosophical dialogues can act as instruments to think together and combine different stakes and problems. The dialogue can provide a shared space to develop a common language where different points of view come together. Next, we explore what a philosophical dialogue is, how it is done and why it might be relevant in the context of ecoliteracy education.

Philosophical dialogues and ecoliteracy

In a philosophical dialogue, participants search for answers to challenging questions under the supervision of a facilitator. The educational use of philosophical dialogues is based on the work of the American philosopher Matthew Lipman who, in the 1960s with Ann Margaret Sharp (Gregory & Laverty, 2018), developed a methodology called "philosophy for children" (Lipman & Sharp, 1988), alongside curriculum resources including stories, lesson plans and exercises in thinking. One example relating to ecoliteracy is *Kio and Gus*, a novel by Matthew Lipman (1982) that explores nature, reality and the difference between saying and doing, and the associated teaching manual *Wondering at the World:* indeed, Lipman and Sharp (1986) argued that the sense of wonder is important in awakening ecological awareness.

In the tradition of philosophy for children, philosophy is not only to be regarded as an academic discipline but as a form of dialogical thinking (Lipman, 2003). Inspired by John Dewey's plea for education focused on the learners (Dewey, 1997), Philosophy for Children aims to teach children to think for themselves (Lipman, Sharp & Oscanyan, 1980; Lipman, 2003). In this approach, the goal is to improve children's critical, creative and caring judgment by having them discuss concepts of importance to them and think about their thinking in doing so (Lipman, 1981). More recently, Hannam and Echeverria have described the deep literacy that comes from philosophical dialogue in a community of inquiry as that which involves "being able to accept and even to become more interested in knowing more about the different ways of being amongst humanity" (2009, p. 94). The skills necessary for collaborative, critical and creative thinking are acquired within the context of language (Trickey & Topping, 2004).

Philosophical questions and dialogues

Can people control nature? Can a mountain have rights? Should access to water be a human right? A philosophical dialogue always involves a big (philosophical) question. Philosophical questions are ultimate questions, "open to informed, rational and honest disagreement" (Floridi, 2013, p. 195). More broadly, questions can be seen as an interruption or intrusion into ongoing processes (Bay-Cheng, 2017). This means that whilst answers are constrained by empirical data and realities, philosophical questions lend themselves to authentic exploration through reasoning and allow us to pause and reflect on the distance between how things are and how they ought to be. This makes philosophical questions useful in exploring key concepts in ecoliteracy such as nature, environment, connectedness and sustainability. These are questions such as:

- Is the quality of life of some people more important than saving the planet?
- Can development be sustainable?
- Is it important to stop climate change?
- Does it matter if words like *acorn, bramble* and *conker* are no longer in school dictionaries?
- Can inconvenient environmental truth be useful?
- What is the relationship between happiness and consumption?

Philosophical questions touch on topics such as knowledge (epistemology), reality (ontology), purpose (teleology) and ethics. Philosophical questions have many possible answers, and the challenge is to identify the better answers and the reasons why some answers are better. In doing so, this allows thinking to be made explicit – what claims are being made and what evidence exists to support these claims, as well as counterevidence and counterclaims? The process of philosophical questioning encourages the development of skills important in the language arts, such as critical reading and comprehension, summary and synthesis, and evaluation of vocabulary choice. Indeed, Hannam and Echeverria (2009) characterise philosophical questioning as that which arises from problematic situations and which leads to the development of enduring understanding and motivation.

Philosophical questions can originate with the teacher or with the student. Students can create their philosophical questions, often in response to a stimulus, which could be the use of a literary text. Discussion of such questions enables students to make connections between the written word and the world. Suitable stimuli for making philosophical questions include the *Poetry and the Environment* collection by the Poetry Foundation (2023), books such as *The Lorax* by Dr Seuss (2012), *The Boy Who Harnessed the Wind* by Bryan Mealer and William Kamkwamba (2016), *Un Lun Dun* by China Miēville (2008) and *Animal's People* by Indra Sinha (2007). Alternatively, images such as those presented in Figures 6.1 and 6.2 can be used to stimulate philosophical dialogue on ecological issues.

In a philosophical dialogue, the facilitator stimulates an investigation of the key concepts in the philosophical questions. Students often sit in a circle and the teacher facilitates the dialogue about the philosophical question by asking further questions (for example: Why do you think so? Are you sure? Can you give an example? or who disagrees?). Students are questioned about the coherence and relevance of arguments and the (hidden) premises or consequences of statements (Rondhuis, 2005). Nurturing reading, speaking and listening in

Figure 6.1 Who speaks for nature? Image credit: Maisy Summer

philosophical dialogue requires attention to language teaching and learning. There is a need for reading support for understanding, critical reading, presentation of information and ideas, listening to and responding to the views of others and text interpretation, analysis and evaluation. Above all, integrating philosophical dialogues into education creates space to relate texts to cultural contexts and local and global issues, and to identify multiple meanings of words and different perspectives on these.

Arguments for the use of philosophical dialogues to develop ecoliteracy

Philosophical dialogues allow students to discuss and explore the meaning of words and concepts. Only by exploring the meaning of the words is it possible to formulate answers to a philosophical question, as students explore the meanings and nuances of (philosophical) concepts and distinct perspectives relevant to the questions. In the context of environmental education, philosophical

Figure 6.2 PhiloZoo cards ((c) Odisee – ExploRatio – De Schrijver, J., Cornelissen, E., Pil, A., Sermeus, J., Van Wichelen, R., 2019)

questions allow the participants in a dialogue to identify problems or challenges in relation to climate change or sustainability, and to make connections between ideas and reality. Furthermore, philosophical questions allow students to bring their understanding and experiences to concepts in ecoliteracy, develop a shared understanding of these concepts and understand how these can be interpreted differently. Philosophical dialogue can also help to identify connections between different parts of a system, and between different systems (for example, economic systems, systems of knowledge and ecosystems).

Studies on the implementation of philosophical inquiries show how these inquiries can be used to help students respond more to each other, identify more problems, ask more questions, make fewer assertions and link thoughts to reasons and experiences more often without prompting (Sprod, 1998). They have also been associated with gains in verbal and non-verbal reasoning and communication (Trickey & Topping, 2007). Large-scale empirical research has also found that attainment in English may well be enhanced – especially for the most disadvantaged students – by doing philosophy for children (Gorard, Siddiqui & See, 2017). Philosophical dialogues support students to reflect critically and develop an understanding of knowledge itself (De Schrijver & Cornelissen, 2023), as words such as "meaning", "truth" or "reality" or "ecology" themselves can themselves be studied in this process. Thus, students can come to an understanding of ideas, the relationships between ideas and the ways such understandings can differ for different people. Examples of categories of talk found in philosophical dialogue include perspective taking, clarification, hypothesising, speculation, definition and differentiation of words, providing examples through the use of analogy, simile and metaphor, introducing new ideas and expressing doubt (Cassidy & Christie, 2013). Beyond the empirical evidence, Wells (1999) draws on Vygotsky's sociocultural theory to argue that thought is not expressed but completed in the word (Vygotsky, 1986), that is, that students must be involved in active co-construction of knowledge through dialogue with each other, their teacher and with texts.

Philosophical dialogues also have the potential to help students to understand and develop argumentation skills. These skills include the ability to develop an inference from premises to conclusion, use deductive, inductive and abductive reasoning, assess the validity and soundness of an argument and identify fallacies or faulty inferences (Baggini & Fosl, 2010). Philosophical dialogues can not only sharpen discussion skills and reasoning ability, but they can also stimulate young people's curiosity and capacity for analysis (Lafortunate, Daniel, Mongeau & Pallascio, 2003; Lipman, 2003). Looking for answers to philosophical questions helps students to develop and evaluate each other's arguments. For example, in examining what actions are most likely to bring about environmental sustainability, a philosophical dialogue might involve an interrogation into the meaning of sustainability, the extent to which environmental sustainability is desirable or sufficient (examining the premises), and evaluation of evidence on different actions such as voting, non-violent direct action, signing petitions and individual lifestyle changes.

Last of all, philosophical dialogues can allow students to engage in co-thinking which requires them to articulate their ideas as words and hold them up to scrutiny by others, and being open to modify or develop their thinking. This act creates discussion about what words (such as veganism, environmental justice and climate crisis) mean and the associations, feelings and attitudes different people have towards them. Through this co-thinking during a philosophical dialogue, Pat Hannam and Eugenio Echeverria (2009) argue that learners are enabled to use their creative imaginations to envision the type of world they want to live in and that they are empowered to participate in decision-making by developing skills essential to democracies such as listening, self-regulation, taking other perspectives into account and responding without resorting to conflict. Through this process, people learn to live together across differences, and also imagine alternative ways of living and being.

Philosophical dialogue about climate and sustainability in educational settings

Two case studies allow us to examine how philosophical dialogue can support the development of ecoliteracy through a focus on the use of language in educational settings.

A climate of uncertainty

The first case of a philosophical dialogue on the topic of climate change is in the context of a series of international youth workshops on uncertainties associated with proposed climate interventions conducted online during the COVID-19 pandemic (Dunlop, Rushton, Atkinson, Cornelissen, De Schrijver, Stadnyk ... & Yuan, 2022). Climate intervention refers to the large-scale intervention in the earth's climate system, by for example reflecting solar radiation into space or by removing carbon dioxide from the atmosphere. At present, these technologies cannot be relied on to significantly contribute to meeting the goals of the Paris Agreement (Lawrence, Schäfer, Muri, Scott, Oschlies, Vaughan ... & Scheffran, 2018), and it is therefore important to be cautious in conversations to avoid raising false hopes – but at the same time to create space to examine the challenges that the prospect of climate interventions raise. The workshops aimed to seek youth perspectives on climate interventions, recognising the need to draw on insights from different disciplines including the sciences, social sciences and arts and humanities.

In the first phase, participants were tasked with conducting research and preparing a presentation on one proposed approach to climate intervention. Language had an important role here, not only in ensuring complex ideas were communicated clearly but also in understanding how words could be used to influence attitudes. For example, framing of climate interventions that invoke associations with nature ("like a volcano" or "like artificial trees") can influence public attitudes (Corner & Pidgeon, 2017), so it was important to identify how language can be used to persuade before tasking participants with the research and presentation task.

In the second step, students were invited to participate in philosophical dialogues. The specific aim of the philosophical dialogue was to identify the right thing to do by exploring the social, ethical and political questions raised by the prospect of different proposals for climate interventions, set against knowledge of the anthropogenic causes of climate change and the known social, political and nature-based responses. Through philosophical dialogues, participants were motivated to think about and discuss key questions underlying the science of climate intervention alongside the political, social and ethical contexts, considering questions like "Can a molecule be bad?", "Does the earth need a doctor?", "Can young people change the world?", "Is science always correct?" and "Are scientists responsible for their inventions?". These questions enabled participants to define the key concepts, identify important structures and systems and the connections between them and discuss different actors and perspectives in discussions on climate change.

In the final phase, participants were challenged to phrase their own questions. These are questions they considered relevant to discuss in the context of climate change. This resulted in questions such as:

- Are socio-political responses better than technoscientific responses?
- Are politicians better placed than scientists to make decisions on climate change?
- Can people reinvent themselves?
- Can people create a healthy relationship with the environment?
- Is it possible to live without harming the planet?
- Is instability a state of human nature?
- Does the environment determine how destructive people are?
- Is destruction of nature part of human nature?
- Can consumer societies be sustainable?
- Is unsustainability part of human nature?
- Can a capitalist society be sustainable?

In this approach, we aimed to work on ecoliteracy education helping students develop the skills to identify and address sticky questions relating social, economic and scientific systems to climate change. As students were providing arguments that often start with "Yes because..." or "No because...", they were invited to think about the meaning of words, language and arguments.

Climate citizens

A second case study took place in Belgian secondary schools. In a project called "Klimaatburgers" (climate citizens), students engage in philosophical dialogues about sustainability and climate science. As discussions in climate science can be controversial, the teaching material aims to explore both the concept of sustainability, the science of climate science and the challenge to develop a shared response to this global problem. This results in dialogic teaching material (Cornelissen & De Schrijver, 2023) focusing on sustainability and climate change and the specific role and nature of science in discussions about climate change.

The method was developed and implemented in Dutch-speaking schools, six groups of secondary school students, and roughly 120 students were involved. This led to dialogues about questions such as:

- Is solving the climate problem more important than maintaining democracy?
- If something is good for one person, is it then also good for everyone else?
- Would it be bad if people went extinct?
- Can one live without harming the earth?
- Are poor people better for the climate than rich people?
- Should rich people contribute more to solve the climate problem than poor?
- Does a scientist know better what is good for us?
- Can we invent something that is good for everyone?
- Can you understand someone who thinks differently than you do?
- Can two people say something else and both be right at the same time?
- How can you know what is good?

Dialogue in action

The transcript presented in Table 6.1 demonstrates what is happening during a philosophical dialogue on climate change.

Students appreciate the dialogic approach. "I like that we can talk about climate for once instead of just hearing about it", a student says. Another student remarks: "Some things really stick, like that question about democracy. And whether a polar bear should have the right to vote. You hear the two extremes and then you find that there's something interesting being said there on both sides". Students observe that this kind of dialogue allows them to explore different perspectives without necessarily having to make a definite choice: "I find it interesting, you hear that people think something different and that they are also right somewhere. or "It's interesting to hear that not everyone thinks alike, but we can still understand each other". Thus, students engage in a process of co-thinking where they learn from each other and develop a multifaceted understanding of climate challenges and the interconnection of the scientific and societal aspects.

Philosophical dialogues invite students to explore the meaning of the concepts that are so often used in the context of climate change, yet are only rarely precisely defined. In the case studies above, concepts like "climate problem" or "democracy" are scrutinised to explicate the meaning of these words. During this explication process, students are invited to provide arguments for their ideas and to adjust their ideas (or arguments) when they change their minds. Not only are (different) meaning(s) of words explored, but also the relationships between concepts are discussed. In the example above, students discuss to what extent democracy and tackling climate change depend on each other. Here, different perspectives are explored. For each perspective a new argument surfaces. And though the point of view of one student may differ from another's, in the end, this dialogic enterprise enables them to see and understand another's perspective. The dialogue does not end with a conclusive answer. Most of the time, philosophical dialogues end with new questions, questions that need to be explored to settle the discussion. This process of co-thinking mirrors a public debate: different voices enrich the question at hand and provide instruments for individual citizens to make choices, at the same time creating a sense of collective identity.

Discussion

Whilst education has enormous potential in helping us to respond to the climate crisis, the danger is that it may serve the status quo through the absence of opportunities to critique systems (linked to petro-pedagogies and obstructive government policies in many jurisdictions) and to imagine and create new ways of being. Although climate and environmental education are often located in science, the language arts are uniquely placed to create spaces for

Table 6.1 Example of a philosophical dialogue where 16-year-olds think about the relation between democracy and climate change

Who is speaking?	*What is said?*	*What is happening?*
Facilitator	*Is preserving democracy more important than addressing the climate problem? What do you think?*	Introducing the big philosophical question
Student A	*That's not a difficult question. The answer is clearly no. Without democracy, life is no longer worth living. Then we are no longer free, but the victims of a dictator.*	Student A gives an answer and argument
Facilitator	*What do the others think?*	Opening up the dialogue to invite different perspectives
Student B	*I agree with student A. Without democracy, we can never solve the climate problem.*	Student B agrees, giving a different reason to student A
Facilitator	*Why do you think that?*	Asking for arguments
Student B	*A dictator thinks only of his own interests, he is never going to choose something that is good for everyone. For that, we need democracy.*	
Facilitator	*Who agrees or disagrees?*	Asking for different opinions to invite extra perspectives
Student C	*I don't agree with that. Now suppose there is a good dictator, surely that is the only person who can help us. Because obviously, democracy is not working well enough, we keep doing nothing, the climate problem is not being addressed.*	Student C provides a counter-reason to the claim that a dictator is never going to think only of his interests
Student D	*Actually, we don't live in a real democracy now anyway. That's the problem. If we had a real democracy, we would be fine. I think preserving or at least building a real democracy is the most important thing to address the climate problem.*	The student asks for concept clarification. What do we mean by a (true) democracy?
Facilitator	*What is a "real" democracy?*	Seeking meaning
Student D	*A society where people consider all voices, and where the power of money or powerful corporations does not always prevail. A democracy where those in power consider the interests of all rather than the few.*	Concept clarification
Student E	*I think the question is wrong. Both are important, both preserving democracy and solving the climate problem.*	Challenging premises

(Continued)

Table 6.1 (Continued)

Who is speaking?	*What is said?*	*What is happening?*
Facilitator	*Why do you think that?*	Asking for arguments
Student E	*You can't have one without the other. Without democracy, you cannot solve the climate problem, because to deal with such a big problem you have to put a lot of people together. At the same time, yes … if we don't solve the climate problem, then eventually there will be no people left because then everyone will die out, and then there will be no democracy anyway.*	The students try to combine multiple perspectives in one answer
Student C	*What do we actually mean by the climate problem?*	Concept clarification
Facilitator	*Good question. Can you help us with that?*	
Student C	*Well, there is global warming.*	
Facilitator	*Is that a problem?*	
Student H	*Yes, because it makes life impossible in the long run. But that still depends. It may be annoying for humans, but there will always be other animals or bacteria for whom it is good. There are bacteria living in hot geysers, after all.*	The student introduces non-human perspectives
Student C	*Whether it is a problem depends on who is looking at it. For polar bears, it might be a bigger problem than for mice. And for humans, yes for humans it's a problem because they have to adapt and organise their world differently. But surely it is also a problem because hm … we have responsibility for other animals. We caused it, so then surely we should care about other organisms, surely they cannot take care of themselves.*	The student introduces the role of perspectives in determining whether something is a problem
Facilitator	*Can't animals take care of themselves?*	New philosophical questions pop up that can be explored further

critique, collaborative meaning-making and exploration which are essential in developing ecoliteracy, and we have identified how this might work in the previous section. Here, we discuss the tensions between dialogue, philosophy and language and identify risks associated with philosophical dialogue on environmental issues and ways of mitigating these risks.

Philosophical dialogue and language

Dialogues depend on (spoken) words. Wittgenstein (1953) famously coined the concept "language games" to describe how words get their meaning in use. It means that the meaning of words is inseparable from the function the words and expressions have and the specific context the words are used in. This implies that to fully understand one another, students need to grasp how context and meaning interact when discussing big questions. For students who are not fluent in the language spoken at school, this can be challenging as they find it harder to understand and respond. Yet, on the other hand, using dialogues can scaffold students' language skills. First, the dialogue stimulates a form of "slow" thinking and talking, a way of speaking in which every word counts and meaning is made explicit, which can be useful for language learners. Furthermore, this slow-down in conversation helps students to focus carefully on the words that are being used. Then, there is the transition between different "language repertoires" where a mix is realised between the everyday use of words, their ecological uses and the connotations these have in public debate. As discussed in the previous section, the meaning of the word "nature" can differ according to the context, and the explanation of the different meanings can provide an opportunity for language learners. The approach described in the previous section provides one model for demonstrating how the teacher can help students discover through dialogue how the meaning of words is built and how in different frames words may differ in meaning.

Philosophical dialogue and knowledge

In a philosophical dialogue, the facilitator takes the Socratic stance. This means that the facilitator does not give any answers, but only questions. This stance involves listening to and drawing out the students' views. There is a mutual responsibility for the development of dialogue, and the facilitator helps the student express and reflect upon their viewpoint by asking questions and identifying inconsistencies in argument. What is challenging in this approach is that the facilitator works with the knowledge students already have (or knowledge that is acquired throughout the dialogue). This means that misconceptions about science or climate change can surface in the dialogues. In this regard, it is important to accompany the activity of philosophical dialogues with activities where students acquire knowledge about the issue at hand. This is illustrated in the first case study (see *a climate of uncertainty,* above). To ensure the dialogue was informed by environmental science, the young people were first informed about the science of climate change and allowed to find out about climate interventions. However, environmental knowledge in itself will never be enough to make political or philosophical choices. In the end, big questions tap into the epistemological or ethical issues that cannot be tackled by empirical knowledge alone.

Indeed, the explicit inclusion of reflection on how language can be used to persuade can help students to reflect on the kind of language that is used in the context of ecoliteracy. In popular writing, expressions such as "It is absolutely true that", "it is a fact that" or "science has proven that" surface. However, these expressions in themselves can be misleading, not only about the substance of the comment but also about the nature of science. After all, science is a discipline where, in principle, any claim can be scrutinised or called into doubt. Understanding science means that one understands that scientific claims can change. Thus, to discuss science in environmental discussions, students also need to master epistemic knowledge and learn to think about philosophical concepts such as truth, tentativeness, causality and uncertainty (Clough & Olson, 2007), at the same time as reflecting that there is a greater than 99% consensus on anthropogenic climate change in the peer-reviewed scientific literature (Lynas, Houlton & Perry, 2021). It is therefore important to consider what might make an interesting philosophical question about knowledge without undermining the current science and harmful climate denialism still prevalent in public communication.

Helping students develop an understanding of (climate) science also means that students step by step learn to use adequate words to describe what scientists do. In this regard, Renée Schwartz (2007) points out we should help students in the context of science to use words such as "valid", "support" or "evidence-based" rather than words such as "proof", "true" or "right/wrong". It is also important to distinguish the different meanings of association and causality, for example in attribution studies which aim to determine whether extreme events can be attributed to climate change (van Oldenborgh, van Der Wiel, Kew, Philip, Otto, Vautard ... & van Aalst, 2021). Precisely by giving room to students to give the meaning of the words they use, they learn to discuss that, for instance, the concept "truth" in science has a specific meaning, that it is not to be understood as an absolute truth, but rather as a kind of knowledge considered true today, knowledge we have good arguments to believe in, yet also knowledge that might (however unlikely it is) change if new evidence is gathered.

The risk of relativism

There is a risk of students implicitly deciding that because philosophical questions can have many possible answers, "anything goes" in terms of a conclusion. However, that is not what we have observed in philosophical dialogues with young people in multiple contexts. Philosophical dialogue must be introduced in a way that suggests that any interpretation of the information is equally as valid as another, but that there are ways of assessing the validity of knowledge claims and soundness of arguments. There must be rigorous and rational debate, which fairly considers the strengths and weaknesses of different positions. Philosophical inquiries open a thinking space that provides safety and helps students explore the nature of knowledge and indeed can help them to learn about the role language plays in society, and the relevance and difficulty in distinguishing rhetoric from truth. This way it can underpin the democratic skills of students and prepare them for an active role in society. It allows them to explore the different perspectives that can exist concerning sustainability and ways to respond to it, and it allows them to create an inner democratic dialogue (Hermans & Bartels, 2020) about complex and multi-angled issues.

The limits of language

Philosophical dialogue involves thinking and talking, yet the climate and ecological crises require urgent action now (IPCC, 2023): should educational endeavours not, therefore, be more focused on action? There are three responses to this. First, thinking and talking are needed in addition to action: it is not an either/or situation. Philosophical dialogue can help different people to identify a shared purpose (teleology) and examine the evidence used to justify actions (epistemology) and the fairness of actions (political philosophy and justice). Second, philosophical dialogue can inform action and provide a space for reflection on different types of action; for example, which actions are most just and the difference between legality and morality. Thirdly, philosophical dialogue can play a role in normalising dialogue around climate change, developing curiosity about the current state of the planet and exploring the potential consequences of different responses. Indeed, countries are building dialogue into national action planning processes; for example, Ireland held a national dialogue on climate action (Government of Ireland, 2022) to facilitate the creation of their climate action to ensure that the actions to reduce carbon emissions were decided fairly, and to empower people to share their perspectives. Thus, thinking and talking are essential in ensuring that action is informed and reflected upon, and that the conversation on climate change is broadened such that there is greater awareness of the scale of the challenge and the possible responses that are already possible.

Conclusion

Often, climate or environmental education is addressed in the science class. This is a necessity, yet not sufficient, for quality climate education and the development of ecoliteracy. To help students explore the meaning of the concepts, connections, perspectives, values and worldviews involved in the societal response to climate change, biodiversity loss and other planetary boundaries, it is relevant to understand the role of the humanities, and in particular the language arts and philosophy. The language arts have a role to play in the creation, analysis, interpretation and critique of literary texts, from poems and books to films, plays and advertisements. These texts can prompt reflection on the perspectives we hold and actions we can take. Philosophical dialogue inspired by these texts can create accessible opportunities for students to question and explore lifestyles, (existential) choices and the relation of humans and their surroundings, all of which are necessary for climate change and environmental sustainability education.

Philosophical dialogue can contribute to ecoliteracy by drawing attention to distinct and shared meanings of related concepts, explicitly focusing on the creation of valid and sound arguments and the creation of space to participate in shared thinking. Whilst we do not argue that every teacher should be trained in philosophy, we suggest that an explicit sensitivity towards philosophical questioning, exploration of meaning and attention to the structure of arguments can help to create the conditions to develop ecoliteracy in the language arts. Thus, rather than teaching students *what* to think, philosophical dialogue teaches them *how* to think, which better prepares them to imagine and create their futures with others with whom they might disagree.

Can words change the world? The question may be as obvious as it is challenging to answer. A word in itself might not suffice; what counts are the actions one takes. Yet without exploring the meaning of the words, we will find it more difficult to articulate shared goals and will not go any further. Studying semantics and argumentation may be an important component for change to occur both in schools and beyond. By helping people study the meaning of words, express ideas in language and reflect on ways to rethink and potentially change the world, we can open perspectives towards a different kind of teaching where students are well prepared to think about global challenges.

Acknowledgments

The Klimaatburgers (climate citizens) project was funded by The King Baudoin Foundation and PWO Research Odisee. A Climate of Uncertainty? was funded by the University of York Economic and Social Research Council Impact Acceleration Account.

References

Baggini, J., & Fosl, P. S. (2010). *The philosopher's toolkit: A compendium of philosophical concepts and methods*. Oxford: Wiley-Blackwell.

Bay-Cheng, S. (2017). What is a question? *ASAP/Journal, 2*(1), 40–42. https://doi.org/10.1353/asa.2017.0009

Berryman, T., & Sauvé, L. (2016). Ruling relationships in sustainable development and education for sustainable development. *The Journal of Environmental Education, 47*(2), 104–117. https://dx.doi.org/10.1080/00958964.2015.1092934

Bravo, A. A., & Chion, A. R. (2017). Language, discourse, argumentation, and science education. In K. S. Taber & B. Akpan (Eds.), *Science education: An international course companion* (pp. 157–166). Rotterdam: Sense Publishers.

Bromhead, H., & Goddard, C. (2023). Applied semantics and climate communication. *Australian Review of Applied Linguistics, 46*(3). https://doi.org/10.1075& /aral.22028.bro

Capra, F. (2007). Sustainable living, ecological literacy, and the breath of life. *Canadian Journal of Environmental Education (CJEE), 12*(1), 9–18. https://api.semanticscholar.org/CorpusID:54512326

Cassidy, C., & Christie, D. (2013). Philosophy with children: Talking, thinking and learning together. *Early Child Development and Care, 183*(8), 1072–1083. http://dx.doi.org/10.1080/03004430.2013.773509

Clough, M. P., & Olson, J. K. (2007). Teaching and assessing the nature of science: An introduction. *Science & Education, 17*, 143–145. http://dx.doi.org/10.1007/s11191-007-9083-9

Cordero, E. C., Centeno, D., & Todd, A. M. (2020). The role of climate change education on individual lifetime carbon emissions. *Plos One, 15*(2), e0206266. https://doi.org/10.1371/journal.pone.0206266

Cornelissen, E., & De Schrijver, J. (2023). *Klimaatburgers*. Odisee: ExploRatio.

Corner, A., & Pidgeon, N. (2017). Like artificial trees? The effect of framing by natural analogy on public perceptions of geoengineering. In A. Maynard & J. Stilgoe (Eds.), *The ethics of nanotechnology, geoengineering, and clean energy* (pp. 361–374). Abingdon and New York, NY: Routledge.

De Schrijver, J., & Cornelissen, E. (2023). Can truth change? Philosophical dialogue to foster pupils' reflection about (mis)information. In L. Parker (Ed.), *Education in the age of misinformation: Philosophical and pedagogical explorations* (pp. 143–161). London, New York, NY and Shanghai: Palgrave Macmillan.

De Schrijver, J., Sermeus, J., & Cornelissen, E. (2018). Is an apple alive? Philosophical dialogue in the science class. ICERI2018 Proceedings. 11th annual International Conference of Education, Research and Innovation, Sevilla, Spain. Publisher: iated. ICERI2018 Proceedings. https://doi.org/10.21125/iceri.2018.0787

Dewey, J. (1997). *Experience and education*. New York, NY: Touchstone.

Dunlop, L., Rushton, E., Atkinson, L., Cornelissen, E., De Schrijver, J., Stadnyk, T., ... & Yuan, X. (2022). Youth co-authorship as public engagement with geoengineering. *International Journal of Science Education, Part B, 12*(1), 60–74. https://doi.org/10.1080/21548455.2022.2027043

Fields, D., & Asbury, K. (2023). Do children think it is important to predict learning and behaviour problems, and do they think genetic screening has a role to play in this? *Journal of Autism and Developmental Disorders*. https://doi.org/10.1007/s10803-023-05966-z

Floridi, L. (2013). What is a philosophical question? *Metaphilosophy, 44*(3), 195–221. https://doi.org/10.1111/meta.12035

Ford, T. H. (2022). Literal and literary atmospheres. In A. Johns-Putra & K. Sultzbach (Eds.), *The Cambridge companion to literature and climate* (pp. 40–54). Cambridge: Cambridge University Press.

Gorard, S., Siddiqui, N., & See, B. H. (2017). Can 'Philosophy for Children' improve primary school attainment? *Journal of Philosophy of Education, 51*(1), 5–22. https://doi.org/10.1111/1467–9752.12227

Government of Ireland. (2022). National dialogue on climate action. Climate conversations 2022. Department of the Environment, Climate and Communications. Retrieved February 13, 2024, from https://www.gov.ie/en/publication/bb1a3-climate-conversations/

Gregory, M. R., & Laverty, M. J. (Eds.). (2018). *In community of inquiry with Ann Margaret sharp: Childhood, philosophy and education*. Abingdon and New York, NY: Routledge.

Hannam, P., & Echeverria, E. (2009). *Philosophy with teenagers: Nurturing a moral imagination for the 21st century*. London, New York, NY and Sydney: Bloomsbury Publishing.

Hecht, S. B. (2021). Domestication, domesticated landscapes, and tropical natures. In U. K. Heise, J. Christensen, & M. Niemann (Eds.), *The Routledge companion to the environmental humanities* (pp. 21–34). Abingdon and New York, NY: Routledge.

Hermans, H. J. M., & Bartels, R. (2020). *Citizenship education and the personalization of democracy*. New York, NY: Routledge.

Hermans, H. J. M., & Dimaggio, G. (2007). Self, identity, and globalization in times of uncertainty: A dialogical analysis. *Review of General Psychology, 11*(1), 31–61. https://psycnet.apa.org/doi/10.1037/1089–2680.11.1.31

Hickman, C., Marks, E., Pihkala, P., Clayton, S., Lewandowski, R. E., Mayall, E. E., Wray, B., Mellor, C., & van Susteren, L. (2021). Climate anxiety in children and young people and their beliefs about government responses to climate change: A global survey. *The Lancet Planetary Health, 5*(12), e863–e873. https://10.1016/S2542-5196(21)00278-3

IPCC. (2023). AR6 synthesis report: Climate change 2023. *IPCC*. Retrieved May 26, 2023, from https://www.ipcc.ch/report/sixth-assessment-report-cycle/

Johns-Putra, A. (2016). Climate change in literature and literary studies: From cli-fi, climate change theater and ecopoetry to ecocriticism and climate change criticism. *WIREs Climate Change, 7*(2), 266–282. https://org/10.1002/wcc.385

Lafortunate, L., Daniel, M. F., Mongeau, P., & Pallascio, R. (2003). Philosophy for children adapted to mathematics: A study of its impact on the evolution of affective factors. *Analytic Teaching, 23*(1), 10–25. https://journal.viterbo.edu/index.php/at/article/view/768

Lawrence, M. G., Schäfer, S., Muri, H., Scott, V., Oschlies, A., Vaughan, N. E., Boucher, O., Schmidt, H., Haywood, J., & Scheffran, J. (2018). Evaluating climate geoengineering proposals in the context of the Paris Agreement temperature goals. *Nature Communications, 9*(1), 1–19. https://doi.org/10.1038/s41467-018-05938-3

Leavenworth, M. L., & Manni, A. (2021). Climate fiction and young learners' thoughts—a dialogue between literature and education. *Environmental Education Research, 27*(5), 727–742. https://10.1080/13504622.2020.1856345

Lipman, M. (1981). Developing philosophies of childhood. *Thinking: The Journal of Philosophy for Children, 2*(3/4), 4–7. https://doi.org/10.5840/thinking1981231

Lipman, M. (1982). *Kio and Gus: Reasoning about nature*. Montclair State University: Institute for the Advancement of Philosophy for Children.

Lipman, M. (2003). *Thinking in education*. Cambridge: Cambridge University Press.

Lipman, M., & Sharp, A. M. (1986). *Wondering at the world: Instructional manual to accompany Kio and Gus*. Washington, DC: Rowman and Littlefield Publishing.

Lipman, M., & Sharp, A. M. (1988). Philosophy for children: A traditional subject in a novel format. *Thinking: The Journal of Philosophy for Children, 7*(4), 32–35. https://doi.org/10.5840/thinking1988748

Lipman, M., Sharp, A. M., & Oscanyan, F. S. (1980). *Philosophy in the classroom*. Philadelphia, PA: Temple University Press.

Lynas, M., Houlton, B. Z., & Perry, S. (2021). Greater than 99% consensus on human caused climate change in the peer-reviewed scientific literature. *Environmental Research Letters, 16*(11), 114005. https://10.1088/1748-9326/ac2966

McBride, B. B., Brewer, C. A., Berkowitz, A. R., & Borrie, W. T. (2021). Environmental literacy, ecological literacy, ecoliteracy: What do we mean and how did we get here? *Ecosphere, 4*(5), 1–20. https://doi.org/10.1890/ES13-00075.1

Mealer, B., & Kamkwamba, W. (2016). *The boy who harnessed the wind*. London: Puffin.

Midgley, M. (1992). Philosophical plumbing. *Royal Institute of Philosophy Supplement, 33*, 139–151.

Miéville, C. (2008). *Un Lun Dun*. London: Pan Macmillan.

Moser, S. C. (2016). Reflections on climate change communication research and practice in the second decade of the 21st century: What more is there to say? *Wiley Interdisciplinary Reviews: Climate Change, 7*(3), 345–369. https://org/10.1002/wcc.403

Nussbaum, M. (1990). *Love's knowledge*. Oxford: Oxford University Press.

Orr, D. W. (1992). *Ecological literacy: Education and the transition to a postmodern world*. Albany, NY: State University of New York Press.

Plato. (2013). *Phaedrus*. https://www.gutenberg.org/files/1636/1636-h/1636-h.htm

Poetry Foundation. (2023). *Poetry and the environment*. https://www.poetryfoundation.org/collections/146462/poetry-and-the-environment

Richardson, K., Steffen, W., Lucht, W., Bendtsen, J., Cornell, S. E., Donges, J. F. ... Rockström, J. (2023). Earth beyond six of nine planetary boundaries. *Science Advances, 9*(37), 1–16. https://doi.org/10.1126/sciadv.adh2458

Rondhuis, N. T. W. (2005). *Philosophical talent: Empirical investigations into philosophical features of adolescents' discourse*. (Doctoral dissertation, Utrecht University). Utrecht University Repository. https://dspace.library.uu.nl/handle/1874/7311

Schwartz, R. (2007). What's in a word? How word choice can develop (Mis) conceptions about the nature of science. *Science Scope, 31*(2), 42–47.

Seuss, Dr. (2012). *The Lorax*. London: HarperCollins UK.

Sinha, I. (2007). *Animal's people*. London: Simon & Schuster Paperbacks.

Splitter, L., & Sharp, A. M. (1995). *Teaching for better thinking*. Melbourne: ACER.

Sprod, T. (1998). "I can change your opinion on that": Social constructivist whole class discussions and their effect on scientific reasoning. *Research in Science Education, 28*, 463–480. https://org/10.1007/BF02461510

Trickey, S., & Topping, K. J. (2004). 'Philosophy for children': A systematic review. *Research Papers in Education, 19*(3), 365–380. https://10.1080/0267152042000248016

Trickey, S., & Topping, K. J. (2007). Collaborative philosophical inquiry for schoolchildren: Cognitive gains at 2-year follow-up. *British Journal of Educational Psychology, 77*(4), 787–796. https://org/10.1348/000709906X10532

UN Economic Commission for Europe. (2012). *Learning for the future: Competences in education for sustainable development*. Retrieved February 13, 2024, from https://unece.org/fileadmin/DAM/env/esd/ESD_Publications/Competences_Publication.pdf

van Oldenborgh, G. J., van Der Wiel, K., Kew, S., Philip, S., Otto, F., Vautard, R., ... & van Aalst, M. (2021). Pathways and pitfalls in extreme event attribution. Climatic Change, 166(13), 1–27. https://org/10.1007/s10584-021-03071-7

Vygotsky, L. (1986). *Thought and language*. Cambridge, MA: MIT Press.

Wallace, H. D. (2019). Transdisciplinary learning in a kitchen garden: Connecting to nature and constructing a path to ecoliteracy? *International Research in Geographical and Environmental Education, 28*(4), 309–323. https://dx.doi.org/10.1080/10382046.2019.1646013

Wells, G. (1999). Language and education: Reconceptualizing education as dialogue. *Annual Review of Applied Linguistics, 19*, 135–155. https://10.1017/S026719059919007x

Wittgenstein, L. (1953). *Philosophical investigations*. Oxford: Basil Blackwell.

Part 3
Creating and responding to texts

7 "Reading-with" to build meaningful connections with animate place-based literacies

Erin Spring and Jennifer MacDonald

Introduction

Children's literature invites young readers to step into fictional, imaginary worlds. In these magical places, inanimate objects often come alive and act as essential characters within the story. For example, there are packs of playing cards that come to life down a rabbit hole (Carroll, 1865/2015); pirates, fairies, mermaids and boys who refuse to grow up (Barrie, 1911/2013); wardrobes that act as portals to a land of talking animals and mythical creatures (Lewis, 1950/2005); and a school governed by wands, seers and wizards (Rowling, 1997). While we have seen how these beloved stories about neverlands and underlands can prompt imaginative play, we wonder: how can we equip readers to attend to the real world – the one that is burning and flooding, desperately needing our care – with the same fascination, in order to generate hope? How might we bring them into dialogue with the local ecologies alive where they live? How might we invite learners to recognise the animacy of life-giving systems that they are connected to?

Setting the context

We write this chapter from the prairie and foothill lands of western Canada. As descendants of European settlers who grew up along shorelines of big lakes in Ontario and later moved west for academic work, we are deeply aware that processes of colonisation privilege our existence and influence our environmental experiences. For example, within our schooling, the dominant narrative of the environment was influenced by the long shadow cast by *terra nullius*. This doctrine, translating to "nobody's land", was used to justify our ancestral settlements, while other violent policies aspiring to erase and assimilate Indigenous cultures and communities were carried out unquestioned. Within our specialisations of Outdoor and Environmental Education (Jennifer) and Children's Literature and Language Arts (Erin), we hear iterations of these stories still being told. Through our careers, we have taught these subjects to all ages – primary, secondary and tertiary. Therefore, the examples we draw upon come from varied experiences within and beyond classrooms, but most

DOI: 10.4324/9781032634524-11

recently teaching pre-service teachers how to engage with children's literature and animate places cross-curricularly.

Rooted in the Canadian imagination is the idea that before European contact, nothing existed. More deeply ingrained in this vision are European-based philosophies of Enlightenment. Born in 17th-century Europe, an emphasis on individualism, progress and anthropocentrism has created the conditions for the school contexts we see today (Donald, 2019). For example, this false narrative of "nobody's land" has been perpetuated in curricula (Chambers, 1999; Donald, 2020): British canonical texts are prioritised; static maps delineate ownership; Jacques Cartier's arrival up what is now known as the St. Lawrence River is the beginning of time; and places are known by Anglo-Saxon names, among others. In children's literature, this means young readers mostly learn the world through colonial, imperialist narratives. While not necessarily compulsory in formal curricula, texts such as R.L. Stevenson's (1883/2016) *Treasure Island*, Daniel Defoe's (1719/2012) *Robinson Crusoe*, Arthur Ransome's (1930/2012) *Swallows and Amazons* and Laura Ingalls Wilder's (1935) *Little House on the Prairie* are often studied in Canadian language arts classrooms and found in libraries and homes. Anecdotally from our experience as educators, within an outdated skills-based programme of study, there is also often limited scope or time for students to question and interpret texts through students' lived identities and experiences in the world. For example, in primary school young readers are taught to interpret texts through skills such as letter identification, phonological awareness and a burgeoning vocabulary. In later years, these practices lead to the reader's ability to trace narrative, plot and character development, climax and setting, often assuming there to be one discernible "reading" that exists *within* the text.

The structure of schooling similarly implies that the most important learning happens when we are inside: sitting in lines of desks, days punctuated by the ringing of bells. Cognitive ways of knowing are foregrounded over experiential or relational practices. Embedded in these ideologies, environments are experienced (and learned) as a *backdrop* to important human activities. Even in some environmental education practices, nature is objectified and is seen as secondary to human experiences. To counter these tendencies, we have learned to foreground place, rather than the environment, in our work. This move prompts us to be critical of the human/environment binary and to focus on the multiplicity of stories (Massey, 2005) within the animate web of more-than-human relationships in which we already live (Donald, 2020; Ingold, 2000). In a time of increased ecological concern, we believe such an approach can open up possibilities for renewed forms of collaboration between the language arts curriculum and elements of environmental education outlined in this book (or, what we will refer to as *place-based learning*).

To do so, our chapter brings forward a process of reading-with. Rather than focusing on magical/fantasy or "other" worlds, we show that the sensibilities to engage with the animacy of place are already embedded in the realities of local place-based ecologies. We begin by articulating theoretical groundings of animacy and place. We then describe how the process of reading-with came to life in other projects to then further provide possible practices to promote relations with animacy. We ask: How might we "read-with" place when we understand places as animate?

Understanding place as animate

Place is a vague concept whose contextuality and subjectivity evade concrete definition. It is much more than a singular or static geographic position but is lived and dynamic. Following cultural geographer Doreen Massey (2005), we see that places lack strict borders, encompass many identities and stories and involve global processes. Throughout her work, Massey draws our attention to the impacts of globalisation (ongoing colonisation), and the influences of the "time-space compression" for how places are perceived and experienced. We worry, however, about the potential influence of how she describes place on the already universalised curricula and the texts used in language arts. Massey's human-centric analysis of place also discounts the life, natural cycles, patterns, stories and knowledges within the *actual* places where learning occurs.

By theorising local topographies, curriculum scholar Cynthia Chambers (1999) helps us to think through these omissions in our understanding of place. She challenges Canadian educators to "write from this place" (p. 144), to develop "a language of our own" (p. 145) and to "build interpretive tools" (p. 146) as part of the difficult work of better understanding ourselves within the uniqueness of geography, but also for enhancing dialogue across differences. Guided by these challenges, and from an Indigenous understanding that prioritises the importance of place, Papaschase nêhiyaw (Cree) scholar Dwayne Donald (2020) articulates that the most urgent concern in education is to honour the unique animacy of places and see them as our living relatives. He writes: "consistent acknowledgement of the animacy of places is not just a neat idea, it honours life and nurtures its continuance" (p. 161).

Animacy expresses a sentience, or the degree of *aliveness* an object contains. In Western scientific worldviews, elements of place are often categorised in ways that objectify them for human use and devalue them as spiritless (Kimmerer, 2013). Consequently, this understanding separates humans and the natural world. In our work, we have learned that there are other ways to experience the natural world such that it is a place full of life and motion. So doing positions humans as part of a living web that has unique patterns based on the ecological and cosmological insights of a specific place. Apalech Tyson Yunkaporta describes the significance of learning to perceive and live by the patterns of an animate place. He writes:

> when you engage with this way of being, you will find it changes you in subtle ways. The patterns of your language will change as you find ways to express the places you come into relation with. Your accent will change as to reflect the landscape you inhabit.
>
> (2020, p. 227)

For us, this means that by centring animacy, and attending carefully to place as alive, we learn to live relationally. Yunkaporta asserts that this shift is a necessary move for becoming "custodial species" (Yunkaporta, 2021, n.p.) that have responsibilities to the whole of the ecological system in which we belong.

As non-Indigenous authors and educators, we understand that we cannot take Indigenous worldviews as our own, but we do wish to learn from such wisdom in order to promote healthier place-based relations, including in our teaching spaces. In order to expand ontological, epistemological and relational conversations with other ways of knowing, we are required to recognise and unlearn the linear and colonial logics of how places are commonly engaged and to generate other frames for noticing, attending and experiencing (see MacDonald, 2020). Anthropologist Tim Ingold (2000) writes extensively about the temporality of place and brings light to the importance of dwelling where one is, in order to read and make meaning. Therefore, we came to see that we do not read *on* or *in* place – rather, to read *with* various ecologically specific life forms. If we understand that place is animate, then we are always reading-with. To explore the interconnections between place, animacy and the possible connections between language arts and environment education, we return to a process of reading-with that we arrived at in previous engagements to more fully consider some of the complex questions that felt unfinished.

Reading-with place

In 2018, we were drawn to the Bow River, which begins in the Rocky Mountains and flows through Calgary before meeting with other rivers that wind through the Prairie, to better understand how we might live, teach and learn responsibly through a series of focused river visits. Chambers (2006) reminded us that "places are like family and friends to whom we are bound by history, memory and love" and that *visiting* place is "a way of renewing and recreating people, places and beings, and their relationships to one another" (pp. 34–35). At the time, we were curious how the river might inform some of the pedagogical conversations we were having around identity and place as non-Indigenous instructors teaching sections of a mandatory Indigenous education course in our faculty of Education (MacDonald & Spring, 2021). Perhaps, we thought, visiting the river would help us more meaningfully bring place into our teaching and learning.

Over several seasons, our visits involved a range of literacy practices, such as letter-writing, journaling, reading alongside and sense mapping. On our last visit, we decided to each select a text inspired by our other river visits to share with each other and the river. After reading the texts aloud, our discussion surfaced with several questions: Were we involving the river in our interactions? How might the river respond or engage with us and the text? How would we ever know, since we do not speak the same language? Through pondering these questions, we realised that, in our process, we had fallen victim to objectifying place, and that we needed to consider how we were *with* place.

We imagined that a process of reading-with might help us shift away from privileging the cognitive intellect (thinking about how a place might respond) to more holistic understandings (the response might surface in unseen but deeply felt ways).

We have carried these wonderings over several years: different students, courses and institutions; new lines of inquiry; moves to different ecologies; multiple expeditions; and personal and professional life transitions. Through time and varied experiences, we now wish to reflect on practices for reading-with in more nuanced ways.

Practices for reading-with as a process of attending to animacy

Through reflexive inquiry (Ellis & Bochner, 2000) we recently came together to discuss what reading-with has taught us while simultaneously practising ideas with respective classes and research collaborators. In generating this chapter, we wish to bring forth possible practices that might help other educators promote or enhance reading-with in their own contexts. In the section below, we offer four practices for reading-with as a process of attending to the animacy of local places and of realising our connections as humans already embedded in a mesh of relationships.

Be guided by local storytellers

To begin, we preface the importance of allowing local stories to guide and conjure. Local stories, both text and oral, need to be sourced and shared to become part of students' lives. In our contexts, we draw on and centre Indigenous children's literature that details stories within an animate place. Instead of inviting children to engage with fictional playing cards and mermaids, or talking clocks, wands, seers, or wizards, these stories often feature sentient beings who provide guidance for living ethically. Additionally, local stories commonly offer depictions of cultural practices that have been shared generationally, thus providing wisdom for living well in a particular place.

For example, we often share with our students Swampy Cree author David Alexander Robertson and Cree-Métis illustrator Julie Flett's picturebook (2021) *On the Trapline*. A boy and his *Moshom* (grandfather) visit their trapline in northern Manitoba where Moshom grew up living on the land, practising *Mushkego* (Swampy Cree) teachings of traditional hunting. Through the book, the vibrancy of the terrain along their route – the animals, colours of the sky and lakes and view of the islands below from the airplane – is evident to the boy. Through their journey, the boy's understandings of the significance of place and of his relationship with his grandfather's experience are transformed. Upon returning to the city, he carries the trapline with him in his memory: "he closes his eyes and can picture the trapline. The trees, the water, and all the land and little islands. Chopping wood and picking berries ... Standing by the lake with Moshum" (n.p.). Engaging this story invites our students to consider their own significant place-relations, and also brings them into a Mushkego worldview.

More locally, we have introduced texts by Siksikasitapi (Blackfoot) authors and illustrators into our classrooms, one of the local knowledges of Mohkinstsis (Calgary). For example, Payne Many Guns' picturebook *Siksikawa Otaawahsinnowa: Where the Blackfoot People Live* (2019) animates the multiple ways in which the Siksikasitapi people live harmoniously with the animals, land and stars on their traditional territory (where Erin currently lives). Its dual-language approach (English and *Siksikái'powahsin*/Blackfoot) prompts readers to engage with this place through a Siksikasitapi worldview in an accessible way. In our teaching, even sourcing local authors who are non-Indigenous provides important history and nuance to illuminate stories and colonial complexities. For example, when learning more about Saskatchewan (where Jennifer now lives), she has read about bird migration and human settlement patterns. One book which has made her see the land differently is William Schroader's *Trees Against the Wind* (2023), where he details that all 350,000 trees on the grasslands – which made settlement possible – have been hand-planted and will only live for three generations on the dry landscape.

Reading–with local texts gives entry points for understanding the multiplicities of places, and for appreciating how the place – and its stories – are *already* alive beyond the text. Educators should take caution, however, that Indigenous stories should not be understood as myth or fairytale. Drawing on a specific story about the Beaver Bundle, Donald reminds us:

> There are many wisdom teachings that can be derived from the story Ksisskstaki Amopistaan. Important insights into Blackfoot imagination and worldview can be discerned by those who are committed enough to engage in careful study and interpretation of its various interconnected teachings. This can take a lifetime. However, it is important to remember that it is not just a story. Ksisskstaki Amopistaan conveys Blackfoot memory, experience, ceremonial ways, and worldview which more generally provide guidance on how to live a good life. So, the spirit and intent of the story itself must inspire people to do something to honour the bundle exchange and the kinship alliance made that day with the birds and animals, or the wisdom it offers becomes static and organic. Nothing grows from it unless people enact its teachings in their daily lives.
>
> (2019, p. 119)

It is vital for our students to understand that cultural teachings are located within the story, and that stories require thoughtful interpretation and patience to understand personal significance and responsibility. Local Indigenous texts offer students in our classes a world view that might differ from their own, but one that nonetheless encourages them to engage with their own place-relations, the power of intergenerational kinships and culturally specific knowledges and stories. In our experience, students may struggle with the interpretation process at first as they expect one correct answer, and seek acknowledgement that their reading is "right". Through continued practice and room for creative ways of expression (for example art, or writing), we find that students grow. Building trust is an essential part of this process: both students and educators need to take risks, practise humility and ask critical questions. Finding ways to understand and honour the teachings within their daily lives takes time and commitment.

Critical questioning to foster relational identity

Relations are impossible to form without studying how humans *contribute to* the well-being of place and how we are also *implicated by* the well-being of that place. For reading-with to occur, there needs to be a critical component to understand the wider societal norms that impact interpretations of stories and places. We are inspired by scholars who advocate for reading as an active, transactional process that necessarily involves an engagement with the world *beyond* the text (Rosenblatt, 1978/1994; Barton & Hamilton, 1998; Cliff Hodges, 2016; Spring, 2018). In order to read-with, focus must be shifted away from text-centric pedagogy (such as book studies, character assessments or plot devices) to attend to *who* and *where* students are in relation to the text. Literacy scholar Margaret Meek (1996) reminds us that "book learning is both life-to-text and text-to-life" (p. 26), highlighting the importance of affective, personal involvement with texts. We must support students in the process of critically considering the implications of who they are in relation to the study of the local place.

Inspired by the work of Gabrielle Cliff Hodges (2010, 2013), Erin worked with a class of twelve-year-olds at a school in rural Alberta. She took up Hodges' *rivers of reading* process, wherein students create collages of their reading histories:

> rivers are constantly in motion and ever-changing, shaped by and shaping the terrain through which they flow … students representing their reading as rivers required them to think about readership not only in terms of the space created by their interconnection with others through reading – families, friends, classmates, authors, fictional characters for example – but also, simultaneously, in terms of time – past, present and, in some cases, future.
>
> (2010, p. 188)

Erin asked the students to map important texts in their reading histories along the shape of a river, representing the colliding trajectories of different moments, places, texts and people in their lives from childhood into adolescence. In so doing, they reflected on questions such as: Where were you when you read specific texts, or heard particular stories? Who were you with? What memories can you recall and what emotions surfaced during the encounter? Would your experience of reading/hearing that text/story have been different somewhere else, or with someone else, or if experienced at another point in your life? In whole-class and one-on-one conversations, students then expanded on these incident collages – offering an opportunity for critical questioning. Several of them had conversations with grandparents or parents about their early-reading memories to fill in some of the gaps. In doing so, the students recognised that memory is not linear: often they returned to their incident collages with new insight and perspective. Others returned with photographs, adding a supplemental layer to the collage. Through this collaborative process – with family and fellow students – they began to think about accessibility, privilege and positioning. Questioning their perceptions, and explicitly having to articulate and map their reading histories, provided students with opportunities to see how their place-relationships influence their reading patterns. Importantly, they were able to see how their identities are forged through overlapping, ongoing relationships with people and places.

It is central to attend to positionality through questions such as: Who am I? Where am I? Where do I come from? Where am I going? In our work with young people, we often begin sessions with introductions that ask students to share parts of their story: where they are coming from, and what brought them to this place. Through questioning, the critical work then begins: How do their stories – similarities and differences – relate to one another and to place? How do their stories inform what they notice? How are power and privilege involved in their stories? What do they think the local place wants them to know? These questions make them an active part of a process that might generate a variety of responses and reactions. Dialoguing about the process – within the process – can support students to uncover something about themselves and their place relations that might promote more ethical connections.

Build language for attending to place relations

Language, when reading-with, is a process of dialoguing with the world. We are mindful, as literary critic Robert Macfarlane documented, that children are losing nature language as technology furthers the clouds of disconnection. Language is important, he writes, because "words act as compass; place-speech serves literally to en-chant the land – to sing it back into being, and to sing one's being back into it" (2015, p. 22). Since language capacity and place-experience are so deeply connected, if we build languages to identify life forms, then we might be better able to form a relationship. For example, when Jennifer moved to Saskatchewan, it was a new ecology for her. At first, everything looked like the same prairie land; however, through attending to birds, she began to recognise the diversity within the landscape and skyscape, and how different species have a unique purpose within the whole. In this vein, we are inspired by Macfarlane's later work with illustrator Jackie Morris (Macfarlane & Morris, 2017) that reintroduces lost words into children's vocabularies and imaginations as a way to awaken the diversity of place and to revive relations.

Further, in *Braiding Sweetgrass*, Potawatomi scholar and botanist Robin Wall Kimmerer says: "to be native to a place we must learn to speak its language" (2013, p. 47). She makes clear that her verb-based language allows the world to come alive differently, compared with noun-based English and scientific ways of interacting with life forms. She writes:

> To me, an experiment is a kind of conversation with plants: I have a question for them, but since we don't speak the same language, I can't ask them directly and they won't answer verbally. But plants can be eloquent in their physical responses and behaviors. Plants answer questions by the way they live, by their responses to change.
>
> (2013, p. 158)

Her sentiments prompt us to not only attend to human-generated taxonomies of language, as guided by Macfarlane and Morris, but to also be aware that life forms have their own languages and expressions. Differing from scientific classification or explanation, there is more than one meaning when engaging place-based ecologies. Thus, honouring places as animate through reading-with comes not only from recognising who our relatives are, but slowing down to observe and listen to how other life forms express themselves and interact with the whole. Both of these processes take time to understand and enact.

In our practice, we have done this in several ways. In Jennifer's Outdoor Education courses, students often undergo a place study assignment and generate their own note-taking system (such as journal entries, poetry, photography, sketching) to attend to what is happening in that place. Each week the students receive prompts intended to build their language. For example, one week they are encouraged to use resources to identify the plants, clouds, birds, winds, among others who are present with the intention to expand vocabularies of place. Then, through the iterative practice of visiting weekly for the duration of the course, they begin to articulate patterns and understand themselves as also storied in the patterns. Since the students all select a different place, and attend in their own ways, their interpretations vary. Similarly, Erin and Jennifer have both brought processes of walking into their pedagogies, wherein students spend time exploring local places on foot. For example, while teaching Susannah Moodie's (1852) *Roughing it in the Bush* at Trent University, Erin's English Literature students walked along the Otonabee River, beyond their classroom walls – the same river depicted in Moodie's text. In doing so, the students experienced the book's descriptions of the river and then met to share observations of their own encounters. In a class blog, students wrote creative pieces, including photographs, maps and letters to Moodie. We also see it as valuable to bring Elders and storytellers to place alongside us to share traditional knowledge. For example, at Nose Hill Park, learning the story of the sleeping buffalo changed how Erin sees the park and how she brings students into the story of where we live and learn in the community.

Together, we have also engaged with Robert Macfarlane and Jackie Morris' (2017) *The Lost Words: A Spell Book*. Their ideas that a picturebook might enliven place-based vocabularies being taken from common use compelled us. Undoubtedly, the text provides a tool for enlivening and engaging with the languages of place and prompting attention that could support articulating stories for a more sustainable world. However, through a process of *métissage* (Chambers, Hasebe-Ludt, Hurren, Leggo & Oberg, 2008) alongside Kainai scholar Tiffany Prete (Apooyak'ii), we questioned how we could take up a British-specific book to dialogue with the complexities of place from our positions and colonial contexts (MacDonald, Prete & Spring, in press). Through our collaboration, the inquiry became less *about* the book and more about the experiences it inspired us to undertake while reading. For example, the book includes words that have been "lost" from British children's vocabularies – such as dandelion, wren, kingfisher and bluebell. While these words and beings exist in this place, many have a deeper story (in other words, are invasive species that were brought through settler contact and influenced native grasslands) that should be heard. Likewise, knowing local languages (stories) about place-based entities can provide important understandings about how to live well in this place.

Through reading-with place, connections with the local community are foregrounded (for example, what do the birds speak?) cultivating deeper relationships. While reading Macfarlane and Morris's *The Lost Words: A Spell Book*, Erin, for example, was prompted to learn the stories behind coulee, drumlin and aspen, words affiliated with her experiences in various ecologies and how they interconnect. Jennifer reflected on a purple crocus she encountered while walking, and recalled a Siksikaitsitapi story about the *ki'piaapi* (Blackfoot for crocus), which provided a deeper meaning around the flower's presence in the spring. Through its call to attend to language, the meaning of the book was not lost, but more importantly, it fell into the background, sparking more attentive relations with the places from which we were reading.

Participate in ecological kinetics and patterns

The word *with*, as part of reading-with, denotes relationality. Since places influence us and we influence places constantly, the *with* involves the uncovering of relationships that might be hidden or overlooked, and intentional participation within the system. Even in learning situations where ecology might not be considered, our connectivity with other life forms is already there (or else humans would cease to exist). Reading-with encourages creative engagement with the kinetics and patterns specific to where the learning occurs. Kinetics here means an act of doing such that "theory and praxis, story and practice are interdependent, cogenerators of knowledge" (Simpson, 2017, p. 20).

In this way, knowledge and knowing are seen more broadly than existing in print-based forms and are generated in the moment of experience. Widening ways of knowing calls us to pay attention to what surfaces holistically through the experience (emotional, spiritual, mental, physical). Learners need opportunities to learn to read, interpret and dialogue with their felt senses in order to realise their connections and to allow for them to grow. Different from reading the winds or clouds, as specified above, this practice is about learning to read our responses in order to see ourselves as part of the ecological community.

As we are constantly moving and adapting within a dynamic system (Yunkaporta, 2020), building dialogical opportunities should centre elements of place. We have found activities most profound when students are brought into conversation with taken-for-granted parts of their experiences. Shadows provide an apt example of a living entity that we often dialogue with on a daily basis without recognition. Shadows are constantly in play with light. David Abram writes about the qualitative dimensions of this life form:

> There do exist a few members of the bodily community that thrive without the dark companionship of shadows – the various winds, for example, or the pane of glass newly set within the window frame. But for most of us material beings, the shadow is part of our makeup .... Shadows are qualitative attributes of the bodies that secrete them. They are time-dependent realms that change their contours with the hour and the season, momentary life zones where the shadow-casting mountain or boulder or body quietly envelops and gathers a range of other bodies under its sway.
>
> (2010, pp. 19–21)

Not many students consider how they are in conversation with entities such as shadows, light or wind, how their bodies produce shadows in relationship to the sun or how their shadows might influence the life of other beings. Pointing out these relationships to students can open an array of stories. Possibilities range from students producing figures with their bodies to telling time based on the position of the sun. In playing with the shapeshifting features of shadows, the goal is not to understand "the how" of these processes, but to participate with them.

In Jennifer's work with secondary school students in outdoor contexts, she uses navigational literacy as a way to disrupt common place-based understandings, and to recognise and enhance kinetic participation on multi-day wilderness trips. Through a mapping process (MacDonald, accepted), students are encouraged to pay attention and to track instances when they feel "fully alive" within the experience. For example, one student examined the feeling he encountered of a cliff watching and protecting him at one of the camps. At the conclusion of the trips, they return to their notes to then map narratives in relation to other entities that were present (such as the sun, wind, rain, clouds, waves, trees, among others) and reflect on unfamiliar responses. Through doing, they began to question how ecological patterns influenced them and how they can find their way through felt-sense as opposed to solely reading a topographic map that is laden with Enlightenment-based ideology. As they noticed these connections more explicitly, they were reading-with through recognising ecological cycles and generating holistic knowledge.

Conclusion

As the consequences of climate change proliferate, language arts and environment education can collaborate to centre healthier place-based relations. In today's contexts, these disciplines are often seen as separate. Place is often represented as abstract, linear and empty. The texts we read privilege worlds other than our own, and it is assumed there to be one discernible "reading" that exists *within* the text. There is often limited room for students to question and interpret texts through their lived identities and experiences. Even in instances where the environment is considered, ideologies underlying literacy often objectify place and make meaningful connections difficult. Following Donald (2020), we can begin by honouring animacy.

To promote the move towards animacy, reading-with is a process that encourages educators and learners to attend to the place-based ecologies and relations in which they are embedded. Some ways in which we have done this are: being guided by local storytellers; critical questioning to foster relational identity; building language for attending to place relations; and participating in ecological kinetics and patterns. While these pedagogical possibilities are not exhaustive, we hope that insights from our reading-with inquiry can inspire others to imagine shifts from reading *on* or *in* place – to rather *with* various life forms – in their contexts.

References

Abram, D. (2010). *Becoming animal: An earthly cosmology.* New York, NY: Pantheon Books.

Barrie, J. M. (1911/2013). *Peter Pan.* London: Puffin Books.

Barton, D., & Hamilton, M. (1998). *Local literacies: Reading and writing in one community.* London: Routledge.

Carroll, L. (1865/2015). *Alice's adventures in wonderland.* London: MacMillan Publishing Co.

Chambers, C. (1999). A topography for Canadian curriculum theory. *Canadian Journal of Education, 24*(2), 137–150. https://doi.org/10.2307/1585924

Chambers, C. (2006). "The land is the best teacher I have ever had": Places as pedagogy for precarious times. *Journal of Curriculum Theorizing*, Spring, 27–38.

Chambers, C. E., Hasebe-Ludt, D. D., Hurren, W., Leggo, C., & Oberg, A. (2008). Métissage: A research praxis. In J. G. Knowles & A. L. Cole (Eds.), *Handbook of the arts in qualitative research: Perspectives, methodologies, examples, and issues* (pp. 141–153). Thousand Oaks, CA: Sage.

Cliff Hodges, G. (2010). Rivers of reading: Using critical incident collages to learn about adolescent readers and their readership. *English in Education, 44*(3), 181–200. https://doi.org/10.1111/j.1754-8845.2010.01072.x

Cliff Hodges, G. (2013). Reading within families: Taking a historical perspective. *Changing English, 20*(2), 182–193. https://doi.org/10.1080/1358684X.2013.788292

Cliff Hodges, G. (2016). *Researching and teaching reading: Developing pedagogy through critical enquiry.* London: Routledge.

Defoe, D. (1719/2012). *Robinson Crusoe.* London: Penguin Classics.

Donald, D. (2019). Homo economicus and forgetful curriculum: Remembering other ways to be a human being. In H. Tomlins-Jahnke, S. D. Styres, S. Lilley, & D. Zinga (Eds.), *Indigenous education: New directions in theory and practice* (pp. 103–125). Edmonton, AB: University of Alberta Press.

Donald, D. (2020). Place. In J. Wearing, M. Ingersoll, C. Deluca, B. Bolden, H. Ogden, & T. M. Christou (Eds.), *Key concepts in curriculum studies: Perspectives on the fundamentals* (pp. 156–166). New York, NY: Routledge.

Ellis, C., & Bochner, A. (2000). Autoethnography, personal narrative, reflexivity: Researcher as subject. In K. Denzin & Y. Lincoln (Eds.), *Handbook of quality research* (3rd ed., pp. 733–768). London: Sage.

Ingalls Wilder, L. (1935). *Little house on the Prairie.* New York, NY: Harper Collins.

Ingold, T. (2000). *The perception of the environment: Essays on livelihood, dwelling and skill.* London: Routledge.

Kimmerer, R. W. (2013). *Braiding sweetgrass.* Minneapolis, MN: Milkweed Editions.

Lewis, C. S. (1950/2005). *The lion, the witch and the wardrobe.* Grand Rapids, MI: ZonderKidz.

MacDonald, J. (2020). A poor curriculum in urban space: An atlas for ethical relationality. In T. Strong-Wilson, C. Ehret, D. Lewkowich, & S. Chang-Kredl (Eds.), *Provoking curriculum encounters across educational experience: New engagements with the curriculum theory archive* (pp. 25–41). New York, NY: Routledge.

MacDonald, J. (accepted). Getting there from here: Mapping as a process to renew kinship. *Canadian Journal of Environmental Education.*

MacDonald, J., & Spring, E. (2021). Learning with the river: Honouring place as non-Indigenous women teaching Indigenous education. In J. MacDonald & J. Markides (Eds.), *Brave work in indigenous education* (pp. 193–204). Lewes, DE: DOI Press.

Macfarlane, R. (2015). *Landmarks.* London: Hamish Hamilton.

Macfarlane, R., & Morris, J. (2017). *The lost words: A spell book.* Toronto: House of Anansi Press.

Many Guns, P. (2019). *Siksikawa Otaawahsinnowa: Where the Blackfoot people live.* Calgary, AB: Durvile and UpRoute Books.

Massey, (2005). *For space.* London: Sage.

Meek, M. (1996). *Information and book learning.* Jackson, MS: Thimble Press.

Moodie, S. (1852). *Roughing it in the bush.* London: Richard Bentley.

Ransome, A. (1930/2012). *Swallows and Amazons.* London: Random House.

Robertson, D. A., & Flett, J. (2021). *On the trapline.* Toronto, ON: Tundra Books.

Rosenblatt, L. (1978/1994). *The reader, the text, the poem: The transactional theory of the literary work.* Carbondale, IL: Southern Illinois University Press.

Rowling, J. K. (1997). *Harry Potter and the Philosopher's Stone.* London: Bloomsbury.

Schroader, W. (2023). *Trees against the wind: The birth of Prairie Shelterbelts.* Regina, Sask: Nature Saskatchewan.

Simpson, L. B. (2017). *As we have always done: Indigenous freedom through radical resistance.* Minneapolis, MN: University of Minnesota Press.

Spring, E. (2018). "Without Manifest, none of the book would have happened": Place, identity, and the positioning of Canadian adolescent readers as literary critics. *Children's Literature in Education*, *49*(2), 101–118. https://doi.org/10.1007/s10583-017-9313-y

Stevenson, R. L. (1883/2016). *Treasure Island*. London: Puffin Classics.

Yunkaporta, T. (2020). *Sand talk: How Indigenous thinking can save the world*. San Francisco, CA: Harper One.

Yunkaporta, T. (2021). Humans as a custodial species. https://garlandmag.com/article/custodial-species/#

8 Environmental education in international contexts

Exploring issues through literature

Hege Emma Rimmereide and Ghazal Shaikh

Introduction

Environmental education is arguably one of the most important fields in education today. Invoked by the UN's Sustainable Development Goals (United Nations, n.d., Agenda 2030, Goal 4.7) and the OECD's *The Future of Education and Skills. Education* 2030 (OECD, 2018), the need to include environmental issues in education is on the international agenda and is receiving increasingly urgent attention. Thus, within the humanities and in the language arts the need to address environmental issues is exceedingly important as it provides a way to create awareness and engagement as well as a means of activating students towards environmental education.

This chapter reports on an international research project involving students in three higher education institutions in England, Norway and Pakistan. The research focuses on engaging students in reading and discussing literature and negotiating meaning. The chapter reports on two cycles of reading and discussion of William Golding's novel *Lord of the Flies* (1954). The choice of Golding's *Lord of the Flies* was founded on the novel being regarded as a classic and being relevant to the curricula of the three universities involved. The novel narrates the story of boys who are stranded on an uninhabited island in the wake of the Second World War. The boys struggle to survive, not only fighting amongst themselves but also having to adapt to life in the rugged natural world.

Embedded in ecocriticism, which explores the relation between literature and the physical world as initially defined by Glotfelty & Fromm (1996) and developed further since, this chapter discusses how students engage in literature conversations about nature in relation to literary texts, or *ecocritical dialogues* (Campagnaro & Goga, 2021; Goga, Guanio-Uluru, Kleveland & Rimmereide, 2023). Ecocritical dialogues emphasise engagement with the topic of the environment and sustainability. Thus, this chapter explores how literature may be a viable way into environmental education within the humanities and explores the role of ecopedagogy and dialogic teaching in education.

The reading project aimed to engage students in transnational discussions. Though the three contexts share some common affinities, there are

DOI: 10.4324/9781032634524-12

also differences. The researchers spent substantial time on agreeing on the setup of the study, discussing readings in relation to the text as well as the prompts for the shared documents where the discussions would take place. Furthermore, it is important in an international project that all participants should be exposed to, and have the opportunity to hold, discussions with peers across contexts. Therefore, the researchers made sure that when setting up the shared space for discussion, all the groups included students from each cross-national context. Following a pre-reading task. the students, still based at first in their respective home contexts, worked in literature circles (Daniels, 2002) which are student-led discussions with designated roles. These in-class discussions were followed by online cross-national discussions with peers from other national groups on Google Documents in response to three prompts set up by the researchers. The Google Documents allowed the students opportunities to communicate with each other despite differences in space and time.

Environmental education through fiction

It is common in school curricula to teach interdisciplinary themes through English (for example: Choo, 2021). Fiction has been used as a tool to discuss citizenship, environment, character education and political ideologies. As argued by Vrasidas, Avraamidou, Theodoridou, Themistokleous & Panaou,u (2015) fiction can provide a space to discuss environmental education in a range of literary contexts (Johns-Putra, 2016).These can include for example climate change and nonfiction children's literature (Campagnaro and Goga, 2021); Black speculative fiction (Nxumalo and Ross, 2019); and young people's fiction (Bigger and Webb, 2010). A number of studies have suggested that ecocritical discussions and dialogues can be facilitated in schools through the medium of literature (Goga et al., 2023; Leavenworth & Manni, 2021 among others).

Different studies have explored fiction and environmental education from varied perspectives. Leavenworth and Manni (2021), for example, have explored the reflections of young learners on nature, climate change, responsibility and sustainability and the discussion of the same topics in three of contemporary Scottish author Julie Bertagna's novels – *Exodus* (2002), *Zenith* (2007) and *Aurora* (2011). They argue that climate fiction can "assist in processes of re-thinking by moving away from real-world causes and instead let readers encounter possible scenarios in a safe space, in this way broadening discussions regarding future sustainability" (Leavenworth & Manni, 2021, p. 739). They suggest that environmental education can be reinforced through ties between literature and education. Goga, Guanio-Uluru, Hallås & Nyrnes (2018) focus on how nature is portrayed in children's and young adults' Nordic literature. They explore analogue and digital texts including picture books, media and story apps highlighting not just Nordic but a global representation of humanity's relationship with nature as represented in children's and young adult literature. Insights from these studies strengthened the possibility of exploring various aspects of environmental education through literature including *Lord of the Flies*, which was the focus of the current study.

Environmental humanities and ecopedagogy

Environmental humanities have evolved in reflexive dialogue with "histories, cultures and values" (Heise, 2017, p. 8). In the 19th century, the romantic poets sought to give voice to nature (Buell, 2005), where notions of wilderness and the pastoral were highlighted. In the 1970s Arne Naess' deep ecology (Naess, 1989) argued that all life on earth has intrinsic value. Posthumanism (Haraway, 2016; Hayles, 1999; Braidotti, 2019) became a focal point in the 1990s, by examining the concept of the "human" in the age of technology and in the age of the Anthropocene (Crutzen & Stoermer, 2000; Heise, 2017; Rimmereide, 2018). The Anthropocene era acknowledges that human impact on the earth has irreversible consequences. Furthermore, animal studies (Derrida, 2002) argue for ethical rights for animals, which further contribute to the importance of plant studies. Increasingly within ecocriticism, new barriers are crossed and new stances are included. In addition to animal and plant studies, Indigenous peoples' thinking and philosophy have gained impetus, and dichotomies such as man-nature, nature-culture, machine-human are being dismantled; for example, Haraway's *Staying with the Trouble: Making Kin in the Chthulucene* (2016) aims to break down such barriers and argues for "kin" as a way to include all species. Concepts such as *wilderness* have attracted critical scrutiny because they are deemed to represent Western-centric ideas (Cronon, 1995).

Environmental humanities emerged from the consciousness of environmental studies and the increasing awareness of the need to eradicate dichotomies between species. Though Heise (2017) points out that anthropocentrism "still characterizes most work in the humanities and social sciences", she argues that "Anthropocentrism itself is a historical phenomenon" (p. 5). Heise acknowledges that the relation between humans and nature has come under pressure and she recognises that there is a move from human agency to ecological agency emphasising the "'species we' implicit in the Anthropocene" (Heise, 2017, p. 5). In this chapter, we attempt to ask how to help students in higher education engage with the environmental humanities.

Ecopedagogical education, or *Ecopedagogy*, is founded on Paolo Freire's seminal work *Pedagogy of the Oppressed* (1970/2005), which centres on critical thinking and dialogue and offers a way to *read the world*. Ecopedagogues internationally draw upon Freire's theories on how reflection and praxis (practice) may develop rights for all. Although environmental teaching practices interact locally, regionally and globally in cultural contexts (Gaard, 2008, 2009; Massey & Bradford, 2011), ecopedagogues emphasise the global approach needed in education concerning the environment. This is particularly important due to the scalar relations between the local and global, or as expressed by Clark, "the challenge of giving sensuous representation of scales in space or time that greatly exceed immediate perception" (Clark, 2019, p. 39). While Garrard (2012) argues that ecopedagogical teaching makes connections to progressive pedagogy in student-centred learning, particularly through cross-curricular activities, which develop active and responsible learners, Kahn (2008) goes further and calls for a more activist stance, which advocates both a practical approach as well as a change of attitude to society in education, particularly focusing on how we view relations between humans and nature. Furthermore, Kahn is concerned with the need for dialogue and solidarity with others with regards to the development of the planet, which requires societal transformations and a need in education to be critical of systems. These changes entail a deconstruction of the hierarchical positions of domination of nature by humans. Since the publication of Kahn's (2010) critical ecopedagogy, which particularly warned against and critiqued neoliberal capitalist development, the Anthropocene has become increasingly more pronounced, challenging fine-grained ecosystems.

Another ecopedagogue deserving of mention in this context is Greta Gaard (2009), who sees ecopedagogy as a way to create awareness of how to teach about environmental matters and who argues that learning happens in, about and through social and natural environments.

Consequently, in an international project, it is important to find a space for dialogue, where participants can express their views and share opinions with each other as a way to enhance *Intercultural Dialogue*, a concept addressed in 2008 by the Council of Europe Ministers of Foreign Affairs (Council of Europe White Paper on Intercultural Dialogue "Living Together As Equals in Dignity", 2008). Intercultural dialogue, which is ingrained in the UN's definition of sustainable development and a response to the need to uphold dialogue across nations, is particularly valuable in education and more so in an international setting across cultural boundaries of space and time. Wegerif's (2007) *dialogic space* encourages the exploration of different perspectives and voices in what he calls an *experiential space* - a concept grounded in, Bakhtin's theory of *dialogism*(1986). The polyphonic character of the dialogues comes through in the multiple voices as reported in this chapter. There are several layers of dialogue in a reading project such as the one presented here. Initially, the participants read *Lord of the Flies* independently. The act of reading literary texts is itself a dialogue between the text and the reader according to Bakhtin's conception of reading (1986). Next, the students discussed the text with their peers, an activity which gave them the opportunity to engage with a range of voices.The dialogues taking place in the Google Documents discussions helped to develop the dialogic space further.

The dialogic aspect of ecopedagogy is further explored by Misiaszek (2020) who states that "ecopedagogues facilitate teacher-student dialogue, questioning benefits and oppression at and between local to planetary spheres, thus *deepening* and *widening* dialogue" (p. 618). Providing a space for dialogue in literature education may encourage students to engage with the literary text and help them make meaning more effectively (VanDerHeide, Newell & Olsen, 2023).

Design of the project

The Sindh, Bergen, York Reading Research Group (SBYRRG) is a team of researchers from Norway, Pakistan and England working on projects regarding undergraduate students' reading of literature. In the two rounds of the fiction reading project under consideration for this chapter, the general design of the project remained the same in both rounds: pre-reading set up for the students, literature circles in class and Google circles internationally. However, there were differences in the readings, roles and probes set up in both rounds.

The readings set for the students in the first round of the project included three articles based on critical analyses of the novel. When selecting these texts, it was important to bear in mind the language proficiency of all the participating students. The readings in the second round were made a little more specific and included an essay on eco-critical analysis of *Lord of the Flies.*

The first round involved general roles in the literature circles including *Discussion Director, Literary Luminary, Word Wizard, Researcher, Scene Setter* and *Connector*. These students were asked to lead the discussion, find textual references, explore the meanings of difficult words in the text, research history and background facts regarding the text, talk about the setting in different scenes in *Lord of the Flies* and connect the novel to their own lives, society and other texts.However, in the second round we decided to include the role of *Nature Scribe*, inspired by Guanio-Uluru's account of reading roles (2019). The task of the Nature Scribe was to focus on the elements related to nature in the novel, including the role of wilderness, plants and animals. The following questions were used to guide the work of the Nature Scribe:

Nature Scribe

A) What role does the use of wilderness play in a) the plot b) characterisation (description of the characters)?

Does wilderness play a symbolic role in the text?

What is the relationship between wilderness and the human characters in the novel?

B) Plant and animals: pay particular attention to how plants and animals are described/presented.

Find samples for discussion. Are plants and animals presented as valuable in themselves or are they only significant relative to their use for humans? What role(s) do plants and animals play in the story?

The Discussion Director role was also designed to enhance opportunities for ecocritical readings:

Discussion Director

> *You will devise at least three higher order questions based on critical readings of the books for the group to discuss.*
> *You need to include a feminist approach, a moral approach and an ecocritical approach – see provided readings.*

The probes in the Google groups were also made more specific; for example from "what elements of the novel did you talk about in your literature circle?" in the first round to "please comment on aspects related to the concepts of nature, democracy and feminism in light of your readings and discussions" in the second round.

Moving towards an environmental design

In the first round (Naylor, McGuinn, Shaikh, Varga & Rimmereide, 2022; Varga, McGuinn, Naylor, Rimmereide & Syed, 2020) where students were given general roles and prompts to engage in their literature circles and Google circles, they responded to various aspects of the novel. The data was analysed in multiple stages guided by Braun and Clarke's (2006) thematic analysis model. Firstly, each team member individually read and coded the data. Then we worked on the data as a group and, following iterative discussions that increased the "inter-observer reliability" (Denscombe, 2007, p. 209), agreed upon these six main codes – social, political, historical, personal, moral and the experience of literature circles.

The analysis of the data from the first round showed that students were able to respond to various aspects of the data including citizenship and ethics. However, there was very little mention of nature in the students' responses. Only two students from England and Norway, out of a total of 39, talked about representation of nature in the novel in one of the Google groups. One of the English students said:

> We also discussed the conflict of man and nature within the book through the destruction of the island at the hands of the boys but also through characters such as Jack, representing the murderous, hunting "man", and Simon, representing nature through his enjoyment of the peace and serenity it bought him.

Responding to this comment, a Norwegian student said:

> I liked your comment about the relationship between man and nature, I hadn't really thought about it! But your comment made me wonder: Are there some characters that are more eco-centric or more anthropocentric in terms of their relationship with nature – what do you think?! [sic] Do they have celebratory relationship with nature or a problematising relationship with nature?

No mention of nature, environment or ecology was made by any other students in any of the other five groups in the first round. The two comments quoted above however are interesting as they indicated that the students possessed the potential to discuss nature-related themes through *Lord of the Flies.* However, they were not given specific instructions including literature circle roles and Google group prompts which might have encouraged them towards a discussion related to nature. There were however some indirect comments about nature. One student for example said: "conch [sic] is the symbol of power". One of the Pakistani students related what happened in the novel to "the mega flood of 2010 in Sindh, Pakistan. How people suffered that time, millions of the people lost their homes and became refugees in their own country". However, in the first round more references were made to moral, historical, political and social elements in the novel.

Keeping in mind the limitations of this first round, the literature circle roles and the prompts in the Google circles were changed. As in the first round's data, we felt that students missed many important themes in the novel including the discussion of nature. We therefore made the instructions in the second round specifically focussed. In the second round, the role of the Nature Scribe and the ecological dimension of the Discussion Director's role – together with more specific direction in the prompts, as discussed above – helped to guide participants to talk specifically about nature in their responses to the novel.

As a result of the changes in approach, it was observed that the students talked about nature more explicitly and more often in this round. Students in the second round made a total of 238 comments regarding nature and its role in the novel, unlike the first round. The data on nature was coded under four main codes. This included the students' discussion of: (i) characters relating to nature (36 mentions); (ii) humans relating to nature (42 mentions); (iii) the presentation of nature in the novel (64 mentions); and (iv) nature as a character (96 mentions). The codes were decided after multiple reiterations within the research team and the meaning of each code was discussed at length before it was agreed. While codes i and ii were fairly clear, the meanings of codes iii and iv needed more discussion and were decided upon in the course of multiple team discussions. When participants discussed the presentation of natural objects such as fire, the conch, the island and so on, for example, these were coded under "the presentation of nature in the novel". However when participants discussed nature as having agency, this was coded under "nature as a character." As one of the English participants said: "Symbolic irony can be seen in the role that fire plays in the novel, first being needed for survival and then being used for destructive purposes later in the book". At points, different team members coded the same code under different categories as in the case of this comment from one of the Pakistani students: "It is shown that until man can not [sic] control the [sic] nature, the nature will destroy him. In order to live in a civilized way human have to control the nature" [sic]. This statement was coded under "nature as a character" as well as "humans relating to nature". However, as more team members coded it as the former code, that was the one agreed upon.

There were different levels of analysis in the students' responses. Some attached symbolic meaning to natural objects. As a Pakistani student commented: "fire symbolises hope". Others, however, observed a change in the presentation of nature as the novel progressed – see for example this comment from a Norwegian student: "The sounds of animals become less noticeable as the story progresses, however, and nature becomes more imposing". Some of the participants were able to make connections between the novel and their own society. As one of the Norwegian participants said: "This supports an ecocritical reading of the book, as it mirrors our own societies' way of treating our planet's flora and fauna". There were a few students, like this English participant, who showed some eco-sensitivity: "I feel that Golding wrote Lord of the Flies [as] a warning of [sic] human nature and the effects we have on the planet and the evil that is within us". An intertextual reference was made by one of the Pakistani students: "nature cares for humans as long as human care for it. This very concept is also represented by Ralph Waldo Emerson's essay 'nature'".

It was observed in the analysis of these findings that though the students demonstrated a capacity to engage with nature and its representation in the novel, they did not indicate any awareness of environmental concepts such as Kahn's (2010) account of critical pedagogy or any discussion of the Anthropocene (Heise, 2017). As noted above, the only intertextual reference made was to Ralph Waldo Emerson. The students recognised that the boys in the novel exploited nature for their own survival. As one of the English students commented:"In the real world, we need to eat for survival, we can just buy our food in the grocery shop, but the boys in book [sic] have to hunt for theirs". However, they did not appear to appreciate the anthropocentricity of their own views, for example when they took natural objects to represent something not for itself but in relation to humans. One of the English students for example said that the conch "was used to draw attention" before adding: "it can also represent rules". One of the Norwegian students commented:

> One may argue that the wilderness possesses some kind of evil, or that human possess something evil that may be seen when human [sic] live in such environments, without any positive role figures [sic]. This is kind of the opposite to what we see today, where people look to the wilderness to "find themselves".

This student picked up on the historical change in perceptions of wilderness, from the Western cultural idea of something wild and dangerous, or "something evil" to the present understanding of finding peace within oneself, in nature, which is currently more a cultural perception (Mellor, 2014). The students perceived wilderness as the Other: "when humans live in such environments" expressing similar ideas to those described by Cronon: "we reproduce the dualism that sets humanity and nature at opposite poles" (p. 11, 1995). Only three of the groups mentioned the word wilderness, despite the fact that the roles guidance specifically invited the students to search for and comment on examples of references to this word. Students in the same Google group commented on the notion of being savage: "The more savage the hunters and the other children get, the more 'savage' and extreme the [sic] nature gets". The students demonstrated somewhat stereotypical notions of wilderness. In this context, all the groups mentioned the word "savage" which may be a way of interpreting and understanding wilderness. Two groups mentioned plants while five groups mentioned animals within this context.

We suggest that a difference in the prompts, instructions and literature circle roles encouraged students to discuss nature explicitly, unlike in the first, less structured round. They were able to comment on different aspects of the role of nature and to take different perspectives on how nature was represented in the novel. However, there was still a lack of criticality that needed to be addressed. We understand that in order to make the students aware of the eco-critical references in the text, we as teachers of literature have an important role to play in their environmental education.

Conclusion and recommendations

From the generalised responses of the students in round one to the more nature-specific responses in round two, we see that the changes to the literature circle roles, Google circle prompts and pre-class readings created a difference in the way students responded to nature in the same text. The students in round one did not talk about nature at all, even though the same novel was being discussed. It appears that since they were not directly asked to mention nature in the prompts and because none of the recommended readings were related to environmental aspects of the novel, the students did not make any comments on the subject. We suggest that this provides evidence that readings and teachers' directions can influence the way students respond to a literary text. Instruction affects the themes students discuss in relation to different texts.

Based on the findings of these two rounds, we recommend that teachers introduce more tools to engage students in a discussion of nature in relation to literary texts. Providing more ecocritical readings of *Lord of the Flies* in the reading list, adding a direct question about nature in the prompts and including more nature-related focus in the roles in the literature circle could help to make the students' responses more nuanced and eco-sensitive. Introducing the students to ecopedagogical metalanguage, relevant contemporary readings and an in-depth understanding of ecopedagogy could help enhance their engagement with environmental understanding through literature.

An international project involving students from different contexts and different experiences involves challenges beyond those discussed in this chapter. Negotiations between different cultures necessitate ethical considerations and a sensitivity to other's perspectives. This may have impacted the students' ability and will to discuss and reveal their opinions as frankly as they might have done. Furthermore, despite the affordances offered by the digital tools used in the project, they may have influenced some of the students' responses. The participants in Pakistan, particularly, generally wrote their comments on their mobile phones, whereas the English and Norwegian students made use of their own personal computers. One advantage of the personal computer is that it allows editing more easily. It might be that these digital issues – together with the fact that the students did not really know the people with whom they were exchanging their ideas – may have inhibited their responses and challenged the concept of a *safe space* (Iversen, 2019). While the students may rely on what they discussed in their groups with peers "at home", the potential lack of a safe space may also have applied to that situation.

References

Bakhtin, M. M. (1986). The problem of speech genres. In C. Emerson & M. Holquist (Eds.), *Speech genres and other late essays* (pp. 60–103). Austin, TX: University of Texas Press.

Bertagna, J. (2002). *Exodus*. London: Picador.

Bertagna, J. (2007). *Zenith*. London: Picador.

Bertagna, J. (2011). *Aurora*. London: Pan Macmillan.

Bigger, S., & Webb, J. (2010). Developing environmental agency and engagement through young people's fiction. *Environmental Education Research, 16*(3–4), 401–414. https://doi.org/10.1080/13504621003613145

Braidotti, R. (2019). *Posthuman knowledge*. Cambridge, Oxford, Boston, MA and New York, NY: Polity Press.

Braun, V. & Clarke, V. (2006). Using thematic analysis in psychology. *Qualitative Research in Psychology, 3*(2), 77–101. https://doi.org/10.1191/1478088706qp063oa.

Buell, L. (2005). *Literary imagination*. Oxford: Blackwell Publishing.

Campagnaro, M., & Goga, N. (2021). Green dialogues and digital collaboration on nonfiction children's literature. *Journal of Literary Education, 4*, 96–114. DOI:10.7203/JLE.4.21019

Choo, S. S. (2021). *Teaching ethics through literature: Igniting the global imagination*. Abingdon and New York, NY: Routledge.

Clark, T. (2019). *The value of ecocriticism*. Cambridge: Cambridge University Press.

Council of Europe. (2008). White paper on intercultural dialogue "Living Together As Equals in Dignity". https://www.coe.int/t/dg4/intercultural/source/white%20paper_final_revised_en.pdf

Cronon, W. (1995). The trouble with wilderness; or, getting back to the wrong nature. In W. Cronon (Ed.), *Uncommon ground: Rethinking the human place in nature* (pp. 69–90) New York, NY: W. W. Norton & Co.

Crutzen, P. J., & Stoermer, E. F. (2000). The 'Anthropocene.' *Global Change Newsletter* 41, pp. 17–18.

Daniels, H. (2002). *Literature circles: Voice and choice in book clubs and reading groups* (2nd ed.). Portland, ME: Stenhouse Publishers.

Denscombe, M. (2007). *The good research guide: For small-scale research projects*. Maidenhead; Berkshire: Open University Press.

Derrida, J. (2002). The animal that therefore I am. In David Wills (Ed.), *Critical Inquiry, 28*(2), 369–418.

Freire, P. (1970/2005). *The pedagogy of the oppressed*. London and New York, NY: The Continuum International Publishing Group.

Gaard, G. (2008). Toward an ecopedagogy of children's environmental literature. Green theory & praxis. *The Journal of Ecopedagogy, 4*(2), 11–24.https://doi:10.3903/gtp.2008.2.3

Gaard, G. (2009). Children's environmental literature: From ecocriticism to ecopedagogy. *Neohelicon , 36*, 321–334. https://doi.org/10.1007/s11059-009-0003-7

Garrard, G. (2012). Introduction. In G. Garrard (Ed.), *Teaching ecocriticism and green cultural studies* (pp. 1–10). London: Palgrave Macmillan.

Glotfelty, C., & Fromm, H. (Eds.). (1996). *The ecocriticism reader: Landmarks in literary ecology*. Athens, GA: University of Georgia Press.

Goga, N., Guanio-Uluru, L., Hallås, B. O., & Nyrnes, A. (2018). *Ecocritical perspectives on children's texts and cultures*. London: Palgrave Macmillan.

Goga, N., Guanio-Uluru, L., Kleveland, K., & Rimmereide, H. E. (2023). Økokritiske dialoger: innganger til arbeid med bærekraft i lærerutdanningene [Ecocritical dialogues: Entries to work with sustainability in teacher education]. Oslo, Norway: Universitetsforlaget.

Golding, W. (1954). *Lord of the flies*. London: Faber & Faber.

Guanio-Uluru, L. (2019). Education for sustainability: Developing ecocritical literature circles in the student teacher classroom. *Discourse and Communication for Sustainable Education, 10*(1), 5–19. https://doi.org/10.2478/dcse-2019-0002

Haraway, D. (2016). *Staying with the trouble: Making kin in the Chthulucene*. Durham, NC: Duke University Press.

Hayles, N. K. (1999). *How we became posthuman: Virtual bodies in cybernetics, literature and informatics*. Chicago, IL: University of Chicago Press.

Heise, U. (2017). Introduction: Planet, species, justice. In U. K. Heise, J. Christensen, & M. Niemann (Eds.), *The Routledge companion to the environmental humanities* (pp. 1–9). London and New York, NY: Routledge.

Iversen, L. L. (2019). From safe spaces to communities of disagreement. *British Journal of Religious Education, 41*(3), 315–326. https://doi.org/10.1080/01416200.2018.1445617

Johns-Putra, A. (2016). Climate change in literature and literary studies: From cli-fi, climate change theater and ecopoetry to ecocriticism and climate change criticism. *Wiley Interdisciplinary Reviews: Climate Change*, *7*(2), 266–282. https://doi.org/10.1002/wcc.385

Kahn, R. (2008). From education for sustainable development to ecopedagogy: Sustaining capitalism or sustaining life? *Green Theory & Praxis*, *4*(1), 1–14.

Kahn, R. V. (2010). *Critical pedagogy, ecoliteracy, & planetary crisis: The ecopedagogy movement* (Vol. 359). Lausanne: Peter Lang.

Leavenworth, M., & Manni, A. (2021). Climate fiction and young learners' thoughts—a dialogue between literature and education. *Environmental Education Research, 27*(5), 727–742. https://doi.org/10.1080/13504622.2020.1856345

Massey, G., & Bradford, C. (2011). Children as ecocitizens: Ecocriticism and environmental texts. In K. Mallan & C. Bradford (Eds.), *Contemporary children's literature and film* (pp. 109–126). Camden: Palgrave.

Mellor, L. (2014). The lure of wilderness. In L. Westling (Ed.), *The Cambridge companion to literature and the environment* (pp. 104–120). Cambridge: Cambridge University Press.

Misiaszek, G. W. (2020). Ecopedagogy: Teaching critical literacies of 'development', 'sustainability', and 'sustainable development'. *Teaching in Higher Education, 25*(5), 615–632. https://doi.org/10.1080/13562517.2019.1586668

Naess, A. (1989). *Ecology, community and life-style* (Trans. David Rothenberg). Cambridge: University Press.

Naylor, A., McGuinn, N., Shaikh, G., Varga, Z., & Rimmereide, H. E. (2022). 'I lost the faith in humanity.' Using William Golding's *Lord of the Flies* to explore ethical and moral issues through transnational, online literary exchange. *Cambridge Journal of Education, 52*(2), 137–153. https://doi.org/10.1080/0305764X.2021.1959519

Nxumalo, F., & Ross, K. M. (2019). Envisioning Black space in environmental education for young children. *Race Ethnicity and Education*, *22*(4), 502–524.

OECD. (2018). *The future of education and skills. Education 2030.* https://www.oecd.org/education/2030/E2030%20Position%20Paper%20(05.04.2018).pdf

Rimmereide, H. E. (2018). 'Who will save us from the Rabbits?' Problematizing nature in the Anthropocene. In N. Goga, L. Guanio-Uluru, B. O. Hallås, & A. Nyrnes (Eds.), *Ecocritical perspectives on children's texts and cultures* (pp. 207–222). Palgrave Macmillan.

United Nations. (n.d.). Sustainable development goals: Agenda 2030. Retrieved January 30, 2021, from https://sdgs.un.org/goals/goal4

VanDerHeide, J., Newell, G. E., & Olsen, A. W. (2023). Conceptualizing dialogic literary argumentation: Inviting students to take a turn in important conversations. *Written Communication*, *40*(2), 417–447. https://doi.org/10.1177/07410883221148680

Varga, Z., McGuinn, N., Naylor, A., Rimmereide, H. E., & Syed, G. K. (2020). We are invited to imagine: Using a literary text to encourage cross-cultural dialogue about citizenship. *Cambridge Journal of Education, 50*(4), 501–519. https://doi.org/10.1080/0305764X.2020.1736002

Vrasidas, C., Avraamidou, L., Theodoridou, K., Themistokleous, S., & Panaou, P. (2015). Science fiction in education: Case studies from classroom implementations. *Educational Media International, 52*(3), 201–215.

Wegerif, R. (2007). *Dialogic education and technology: Expanding the space of learning.* Berlin, Heidelberg, Dordrecht and New York, NY: Springer.

9 Pre-service English teachers engage with the natural world through poetry

A cubed template to document the plains of the unknown world

Amanda Naylor and Nicholas McGuinn

Introduction: setting the scene

What part might poetry play in the language arts' contribution to environmental education? How might poets model ways of responding to and interacting with the non-human world through their work? These questions take on a particular resonance, perhaps, when the poets in question are pre-service teachers studying to become secondary English teachers and the audience for their poetry consists of the young people they are preparing to teach – the same young people who, far into the 21st century, will have to bear the brunt of an ecological crisis not of their making.

To begin to answer those questions, we turn to Adeline Johns-Putra and Kelly Sultzbach's suggestion that climate can be categorised in terms of "three distinct but interlocking definitions of index, agency, and system", where *index* might refer to a "data-driven patterning of weather" (2022, p. 2). Scientific research has produced a formidable amount of information regarding the current condition, not only of the climate, but of the planet as a whole. But by whom and how and to what end is this information received? Dominic Head cites Dipesh Chakrabarty's account of *doubling* to describe "a new duality of human consciousness" (2022, p. 2) – it may be that our intentions towards the environment might, on the one hand, be considered as fundamentally benign; but on the other we are also "obliged to see ourselves as part of an 'impersonal and unconscious geophysical force, the consequence of collective human activity'" (cited in Head, 2022, p. 2). Such self-knowledge could provoke a range of negative responses: fatalism, guilt, a sense of powerlessness. As the poet Marcella Durand writes: "At times it seems as though everything we touch, every desire, every process of life we have, turns nature to dust" (Durand, 2011). Added to these woes might be a feeling of bewilderment at the complexity of the information that climate scientists need to share. How is one to comprehend, for example, the vast differences of scale between a human lifespan and deep geological time? Or make sense in human terms of this bleak statistic from the World Bank:

DOI: 10.4324/9781032634524-13

> In 2019, high-income countries were responsible for over 40% of global consumption-based CO2 emissions, while the contribution of low-income countries is a negligible 0.4%.
>
> (World Bank, 2023)

Ann Fisher-Wirth and Laura-Gray Street, editors of a major ecopoetry anthology published a decade ago, preface their collection with the disclosure that "we've become ever more convinced that the environmental crisis is made possible by a profound failure of the imagination" (2013, p. xxvii). For the poet Jorie Graham, one of the contributors to their ecopoetry anthology, it is the emotions which provide a potent means of addressing that failure:

> They [the general public] feel they "know this [scientific data about climate change] information already, so why do they need it in a poem." That is precisely the point. They "know" it. They are not "feeling it." That is what activists in the environmental movement are asking of us: help it to be felt, help it to be imagined.
>
> (Graham, 2012, p. 40)

It has become a truism to say that poets have engaged with the non-human world for millennia; but is there a specific kind of poetry which is particularly suited to the task described here by Graham? Fisher-Wirth and Street's *Ecopoetry Anthology,* comprising more than 600 pages and featuring over 200 poets, takes a broad and inclusive approach to the term which, in its *historical* section, finds room for such established figures as Walt Whitman and T.S. Eliot. Others, however, would insist on a far less catholic definition of ecopoetry. Lynn Keller, for example, dates a marked change of perception to the publication in 2000 of Paul Crutzen and Eugene Stoermer's brief but influential paper entitled *The Anthropocene*, which, she argues, encouraged an awareness that we live in a "period of *changed recognition*" [original italics] and have therefore become self-consciously aware of "the responsibility humans bear for the condition of the planet". Keller believes it is possible to identify a specific poetry belonging to what she calls "the self-conscious Anthropocene" a poetry which "responds to contemporary environmental changes and challenges" (2017, p. 2). Jonathan Bate once described poems as "imaginary parks" (2000, p. 64) where we might find refuge and consolation from the pressures of an urbanised and industrialised world. For Keller and those who share her views, the current environmental situation means that a poetry focused upon what might be termed *retreat narratives* (Head, 2022) will no longer suffice (if it ever did). Juliana Spahr cites her fellow poet Rob Wilson's coinage of the term *747 poems* to describe poetry written by tourists holidaying in Hawai‘i. They expend substantial amounts of fossil fuel in order to jet in for their nature fix and then jet out again without any real sense of engagement with an environment which they have used as just another commodity designed for their (guilt-free) gratification. "I was more suspicious of nature poetry", Spahr writes, because:

> even when it got the birds and the plants and the animals right it tended to show the beautiful bird but not so often the bulldozer off to the side that was destroying the bird's habitat. And it wasn't talking about how the bird, often a bird which had arrived recently from somewhere else, interacted with and changed the larger system of this small part of the world we live in and on.
>
> (2011, p. 69)

But even if a poet writing from the perspective of the self-conscious Anthropocene were to acknowledge the presence of Spahr's bulldozer and the human power structures which mobilise it, that would still not be enough. Timothy Clark notes an "intellectual tameness" in "much recent poetry written explicitly on the topic of global warning". Such pieces, he writes:

> tend to be lyric poems, in a first-person voice, which may well express a powerful sense of protest and loss but yet do little to challenge inherited Romantic assumptions that pit human affairs against a would-be sacralised "nature".
>
> (2019, p. 59)

An alternative poetic approach might be to encourage a concerted effort to challenge that relentlessly individualistic and anthropocentric gaze so that other, non-human perspectives are foregrounded in poetry, and the living beings to whom those perspectives belong are, in terms of environmental justice, accorded their rights as sharers of the planet. Clark cautions against the temptation to believe that poetry – on account, for example, of the associations between rhythm and metre, the beating heart, the menstrual cycle or the seasons – is uniquely well-placed for this task because its language is somehow "more 'natural' or redemptively 'organic'" than other forms of linguistic expression (2019, p. 68). On the contrary, he argues, poetry's true potential lies in its capacity for "carefully judged formal inventiveness" (2019, p. 75) – allusive, many-layered imagery, rhetorical strategies, aesthetic crafting, the reconfiguration of traditional verse forms – to open up "the space between the human and the non-human" where, Clark argues, "poetry occurs" (2019, p. 61). At the opposite end of the ecopoetic spectrum from those poems written in lyric form for individual speakers to voice their sense of protest and loss lie the works of poets who have taken to an extreme the challenge of opening up "the space between the human and the non-human", not only by interrogating the sense-bearing capacity of human language but also by bringing the sounds and communication patterns of the non-human world into their poetry. To cite just a few examples: Evelyn Reilly's collection *Styrofoam* (2009) incorporates chemical formulae into the poems; Jody Gladding's *Translations from Bark Beetle* (2014) "carries the language of bark beetle from beneath the bark of the tree, across to a rubbing, across into human language" (Moe, 2018, p. 451); a. rawlings' *Wide Slumber for Lepidopterists* (2006) includes images and shape poems which simulate the sound and structure of butterfly wings. In works such as these, "[l]anguage and thought become defamiliarised precisely as they move towards experiences beyond the human scales and norms which they usually express" (Clark, 2019, p. 62) through "an attentiveness to the poesis of another species" in what Aaron Moe calls "zoopoetics" (201, p. 451).

Keller's vision of a poetry for the *self-conscious Anthropocene* is not unproblematic. "Part of theatre's function in the age of climate change", Theresa J. May writes, "is grief work" (2022, p. 135). It is appropriate that we have the opportunity, through the arts – and not least through poetry – to mourn the non-human world which is disappearing before our eyes and which we have done so much to destroy. However, as Evelyn Reilly argues in her essay *The Grief of Ecopoetics* (2011):

> if you combine grief over our limited role as citizen-artists with environmental grief the place from which you are writing can get almost paralyzingly grim.

In seeking a more positive way forward, Reilly draws a distinction between the direct action which is the "common responsibility" of all citizens and the kind of agentive response which might be expected of poets. In words reminiscent of W.B. Yeats' poem *Lapis Lazuli*, she declares:

> if you … think, as I do … that there is a kind of power that comes through art, then even in the grip of overwhelming grief, the key may be to find a way to still work out of the joy and aesthetic pleasure that are essential to that kind of power.
>
> (Reilly, 2011)

If ecopoetry continues to push at the limits of poetic lexis, syntax and form, it risks alienating its potential lay audience and, as Clark warns, confining itself to an academic silo. If it focuses relentlessly on grief, guilt and disaster, it risks inculcating in its potential audience that sense of grim paralysis described by Reilly. Before considering how these issues did or did not play out in the poems written by our pre-service teachers, we will now provide some contextual information about the project itself.

Account of the project

The project required that arts students training to be teachers of English to native speakers in the North of England were asked to engage with discourses usually related to science. In seeking to liaise with science colleagues, we were mindful of Ursula Heise's observation that collaboration between scientists and environmental humanists – in whose number we included ourselves as English specialists – is "indispensable" (2017/2021, p. 2). The project was a practical science fieldwork trip for 36 pre-service teachers of science and

English. The project was nested within a larger European Erasmus+ project which was focused on developing teachers' use and understanding of mobile technology within educational contexts. One of the arguments within the literature on teaching with mobile technology is that it enables a more authentic experience for learners, as they are able to work outside of the classroom and with a relative amount of autonomy (Burden & Naylor, 2020). In particular, this project focused on developing collaborations between English and science pre-service teachers, working together to produce an e-book based on tasks undertaken on a field trip to a North-East Coast of England seashore. The choice of a coastal site for our collaborative work offered opportunities to help us develop our engagement with a *Blue Humanities* focus on environmental thinking (Mentz, 2024).

The specific scientific task for the pre-service teachers on the field trip was to research the way of life and habitat of one alga and one animal found upon two seashores, one exposed to and one sheltered from the North Sea winds. In making this choice, we were trying to avoid the temptation of creating a taxonomical hierarchy of non-human creatures, in which those which most resemble or appeal aesthetically to human beings are given privileged access to environmental resources and are most prone to anthropomorphic sentimentalisation (Keller, 2017). The pre-service teachers were tasked to work with observation, to acknowledge and respect nature through counting and undertaking scientific data collection. The science and English pre-service teachers used their mobile devices to record images, sounds and video clips of their experiences. They applied scientific principles to count and observe the flora and fauna using the *belt transect* method, whereby an examination of an area where there is a gradual change from one side of the habitat to another is undertaken. Using a quadrat, they identified and counted the species living within the measured square. They worked on various points on two shores. Most of the scientists found this field work within their comfort zones, whilst for the English pre-service teachers this was a definite challenge and more or less completely new. It seemed that the English graduates were very unfamiliar with the practical aspects of scientific study and their responses to the work varied widely.

During the day of the field trip and data collection, the English pre-service teachers, in addition to working on the scientific data collection, were tasked to use their mobile devices to collect details and impressions of the seashore throughout the day. They were to photograph, film or record their sense impressions, the setting, the organisms and their feelings about the day, as the basis for later writing poems in response to the day's work. The medium of writing in poetry was chosen for the project because it requires detailed attention, as Clark (2019) points out, to crafting language, which was a significant skill to focus on developing with the pre-service teachers. Also, the writing of poetry is often marginalised in both school and higher education contexts (Dymoke, Lambirth & Wilson, 2013), so this we regarded as a valuable element of the project.

The creative outcome of the project was to be an e-book for school students around the age of 11–12 years, to supplement their science or English curriculum resources. The e-book was to contain information about the context for the trip: video clips, sounds of the seashore, photographs of the organisms that could be found there, as well as information gathered for the scientific task itself. The e-book was also to include poems written by the pre-service English teachers about the trip and aimed specifically at this age group. In writing for this book, the English pre-service teachers were asked to engage imaginatively with the scientific data that they had collected and observed on the field trip and use this as a basis for their poetic writing.

After the field trip, the scientists analysed the data they had collected and produced kite diagrams showing the abundance and distribution of their chosen organisms on both the exposed and sheltered seashore. Population estimates of limpets and sea anemones were also calculated and displayed in the e-books as bar charts or pie diagrams. Additional information about the organisms' way of life was added.

The English pre-service teachers had a follow-up creative writing session to create their poems. As is often the case with writing poetry, the pre-service teachers were daunted by the task (Dymoke, 2003). The brief for this was fairly open – they were asked to use stanzas and not to produce rhyming poems (Brownjohn, 1980). These guidelines were implemented as a way to structure how the pre-service teachers wrote, along with paying attention to what might appeal to the younger, Key Stage 3 (11 to 14 years) students. Furthermore, it is argued that for inexperienced writers, searching for rhyming vocabulary limits the width of expression available for poetic expression. Lastly, the pre-service teachers were encouraged to think about the field trip from the point of view of the inhabitants of the seashore. These guidelines were suggested rather than definitively required, because we wanted the pre-service teachers not to feel limited in their responses, rather to open up their writing. It is very interesting, therefore, to see in their responses the variety of representations of an anthropocentric and a non-human perspective. Having produced their initial drafts, the pre-service teachers then had another workshop with a published writer of children's poetry who was also a former teacher, to take into account the audience of young readers for their poems and to work with them to edit their texts so that they might be accessible for school students aged 11–12, who were the audience for the e-books. It is these poems that we analyse in the next section.

In order to focus our analysis of the poems on the ways that the pre-service teachers had positioned themselves in relation to the environment in their writing, we used Buell's (1996) definition of an environmentally orientated text to guide our analysis. The principles are (i) that "the nonhuman environment is present not merely as a framing device but as a presence that begins to suggest that human history is implicated in natural history"; (ii) that "the human interest is not understood to be the only legitimate interest"; (iii) that "human accountability to the environment is part of the text's ethical orientation";

and (iv) that "some sense of the environment as a process rather than as a constant or given is at least implicit in the text" (pp. 7–8). We felt that using this framework, along with our professionally developed poetry analysis skills, would help us to gain insight into the pre-service teachers' responses to the interdisciplinary project. When we looked through different ways of exploring ecopoetry, Buell's criteria appealed to us because they seemed to offer a convincing and systematic starting point for exploring the pre-service teachers' poems from an environmental orientation.

Exploration of pre-service teachers' poetic responses

As suggested earlier, our decision to locate the poetry-writing project on the shoreline of the North Sea offered our pre-service teachers an opportunity to experience the "contrast between wet swirl and dry structure" which, Steve Mentz suggests, "operates as a fundamental division that animates blue humanities thinking" (2024, p. xiv). Given that the content of our pre-service teachers' degrees contained a substantial amount of English Literature, we were particularly struck by Mentz's proposition that "[l]iterary culture" might act as a "bridge-discourse to make water-infused ideas accessible for thinkers in many modes" (2024, p. xv). How would our pre-service teachers conceptualise the shoreline in their poetry? Would they attempt to describe the actual physical location, or would they take the *idea* of a shoreline as a stimulus for a more free-ranging, imaginative engagement?

In this context, it is interesting to compare three poems which share the same "dry structure" focus: the chalk cliffs fronting the actual shoreline accessed on the visit. Nicola (all names have been anonymised) called her poem *Headland* and, as that straightforward title suggests, attempted a metaphorically and alliteratively rich description of what she saw:

> Land's end looms
> Then drops –
> Eroded, decayed incisor
> Where salt surge lashed.

Like Nicola, the speaker in Rebecca's poem is positioned as an observer who describes what they see but does not intervene. However, Rebecca's choice of title, *The Cliffs Peered Down*, suggests something more than an attempt to describe what the speaker observes. Rebecca's chalk cliffs become the "white faces" of powerful beings who sit in judgement on humans and animals:

> The cliffs peered down upon the souls,
> Of those who dared to pass through their archways.
> No beast, or man, could match their reign,
> Those dominating white faces judging.

Rebecca's speaker might not intervene in the poem, but they possess a knowledge of the workings of deep geological time which enables them to bear ironic witness to the cliff's own vulnerability. The speaker understands that the cliffs themselves will eventually suffer the same fate as those they now judge:

> Until time will come, they can judge no more,
> Because she has washed away their scornful eyes.

The third of these *cliff poems* is *Intrusion,* by Jessica. As the title might suggest, it takes a very different stance from the other two, not least in that it explores Mentz's "contrast between wet swirl and dry structure" (2024, p. xiv). The location is still the site of the actual visit: the pre-service teachers are shown exploring the shoreline, as they were asked to do. However, the perspective has shifted radically: the speaker is not some detached human observer but the sea itself, outraged at this "intrusion" into her space and her family:

> Traversing down the treacherous path,
> I see them coming with their weapons of science, ready to trample and intrude
> Upon my peaceful solitude
> I lie bare before them, waiting.
> How can one measure my beauty?
> Create patterns and diversity among my children

If the sea in this poem is gendered as a mother, then the cliffs are – not judges this time – but the sea's "fathers" [sic] who must watch helplessly as their daughter's world is subjected to "chaos and upheaval" at the hands of the human intruders:

> You ask, 'why do I allow this violence?'
> The cliffs, my fathers, look on
> As though nervously waiting for my torment to end.

Timothy Clark reminds us that:

> [O]ur perception and understandings of the non-human is [sic] always mediated, constrained by the finitude of our own nature and its particular, necessarily anthropocentric take on reality.
>
> (2019, p. 60)

Each of the three *cliff poems* is heavily anthropocentric in its engagement with the non-human world: Nicola's cliffs "square up" or "stand resolute"; Rebecca's "judge" or "fail to notice"; Jessica's "look on … nervously". While all three are set on the shoreline explored during the actual visit, the sites which feature in the poems by Rebecca and Jessica are starting to transmute into something less rooted in reality and more open to imaginative interpretation. Lines like Rebecca's "The cliffs peered down upon the souls,/ Of those who dared to pass through their archways" brings to mind a scene from a fantasy novel like J.R.R. Tolkien's *Lord of the Rings*, where the travellers pass through a mountain gorge guarded by gigantic statues. Elsewhere in the poem, Rebecca adds a line which could have come straight from the Bible's *Book of Psalms*: "They cast their stones on the path of the righteous". The speaker in Jessica's poem might be a Nereid from Greek mythology.

The more we reflected upon our pre-service teachers' poems, the more it seemed to us that our English Literature graduates' perceptions of the North Sea shoreline were – as the references to *The Cliffs Peered Down* and *Intrusion* suggest – heavily influenced by anthropomorphic cultural tropes. This could be seen, for example, in the title of Tim's poem which he called *Shore Shank Creatures* in punning allusion to the film *The Shawshank Redemption*. A number of the pre-service teachers chose, like Jessica, to write from the perspective of the sea, the shore or one of the creatures who live there. Even when trying to adopt this point of view, however, they recast the experience in human terms which relied upon literary and cultural allusions to achieve their effect. For example, Larry's piece *Attrition* – another interesting title – purports to be a dialogue between sea and shore but in effect is a poem about sexual seduction which bears no small resemblance to something that might have been written by John Donne or Andrew Marvell:

> I am shore
> To invite you, entice you.
> Come, sup with me …
> … I am deep,
> Daring to drive and breach
> Your abandoned spirit.

Keely's poem *Ocean's Bounty* establishes a conceit in which seaweed is described as the child of the sea; but in what reads like a twist upon the Greek legend of Hades, Demeter and Persephone, this particular daughter wants to escape from, not be reunited with, her mother. The wind murmurs to the seaweed:

> … how it will be welcomed back to the loving, enfolding arms of
> The mother far beyond the shore.
> But the seaweed resists with exhausted, feeble waves.

Clark notes how a poem like Ted Hughes' *Hawk Roosting* – which also purports to represent a non-human perspective – attempts to challenge the anthropomorphic gaze by "having the bird express values that, in human terms, would be outrageous or insane" (2019, p. 63). Whether consciously crafted or not, this is certainly a strategy employed in a number of our pre-service teachers' poems. The further the depictions of the shoreline move away from the actual site of the visit, the more they are described as some kind of existential battleground where the human and the non-human meet in combat. Imogen's poem *The Heckled Hermit* is written from the point of view of a hermit crab which clearly has no time for the human beings who disturb its habitat:

> … my home is invaded
> by those intrusive outsiders and their imposing ways …
> … The sea offers no place for them – She is the mistress of our kingdom
> …
> … She'd swallow them whole and spit them out on the jagged rocks …

Words like these, supposedly issuing from the mouth of a hermit crab, would have a very different resonance if spoken by a human being. "[W]ater-infused ideas" (Mentz, 2024, p. xv) can take many forms, not all of them positive.

Science and the space between human and non-human

The poetry of the pre-service teachers represented their ways of looking at the seashore, shaped by their disciplinary training as arts graduates. However, the brief of the project was that they worked with scientists and utilised scientific methods and equipment. The poetry produced by the pre-service teachers varied widely in their responses to the scientific task, ranging from making no reference to the science at all, simply using some scientific terminology in their writing or responding imaginatively to the scientific work and artefacts that they were using.

Across the poems, a whole range of attitudes to the science of the task are manifested. Scientific terms are used to make observational descriptions more specific and to convey impact, such as "aileron" (Nicola), "littoral land" (Larry) and "Chondrus Crispus" (Imogen). In Jessica's poem, *Intrusion*, where the speaker is the sea, the scientific tools that the humans are using are depicted as "weapons of science", threatening the peace and solitude of the environment. In Tim's poem *Shore Shrank Creatures*, the speaker catalogues what he sees on the sea-shore from his human perspective:

> Shore shrank sea creatures,
> sparkling wash of land's crooked hand,
> calloused, rustling earth loose under foot
> ever-shifting, ever-changing
> Torturous dead ivy like snaking queues of schoolchildren,
> A cubed template to document the plains
> of the unknown world

From that knowing pun in the title to using anthropocentric terms such as "hand" and "foot" to characterise the shore, the speaker sees everything in human terms. However, undertaking a scientific task and using the quadrat has resulted in Tim creating – whether consciously or not – a very powerful, almost Blakean metaphor: one thinks of Blake's picture depicting Newton measuring the universe. He depicts how he as a scientist is using the equipment to count the organisms. But the quadrat takes on a metaphorical role to "document" what is "unknown", as if to colonise a world that is not known to the speaker. In his role as a scientist he has the power to regulate and organise that world: the non-human, known or unknown, is naturally subject to him, a human being. There is no sense in this poem of Tim questioning this hierarchy or extending imaginative empathy to the non-human.

Two poems in particular convey how the task had impacted on the pre-service teachers' vision of the non-human, in line with Clark's (2019) identification of the possibilities of poetry opening spaces between human and non-human. Both of the writers quoted below describe the use of the quadrat for counting and estimating, shaping their visual perceptions and how they responded to what they observed. *Shore-footedness* opens with:

> Trepidatiously [sic] he takes gigantic welly steps
> On a transect from sea to sea-shore
> Angling for a look at the ecosystem
> Squares focused like a camera viewfinder

Daniel foregrounds the feeling of the speaker as unsure, somewhat unstable in the rubber boots or "wellies", that he has had to don for the scientific exploration of the seashore. Using the term "transect", directly pertaining to the task of examining where one habitat (sea) changes to another (sea-shore), he then portrays his actions as "angling" to see the ecosystem and specifically references the quadrat, choosing the less formal term of "square" to characterise how the quadrat has worked to shape his vision, in the way that a "camera viewfinder" would. The next stanza describes:

> The phenomena of organisms organised in grids.
> Perpendicular is the pursuit, perpetrated
> By amateur mariner scooping for a look,
> Sucked in by colours red, amber, green

Again the feeling of trepidation is present: he is an "amateur mariner" (not confident in his interactions with the marine environment) but the scientific approach using the quadrat guides him and has the effect of taming and controlling the organisms for him to survey – their world is now in "grids" – subject to human control, and the speaker is depicted as pursuing the organisms, "scooping" into their world, attracted by their colours and what they might offer his "perpendicular", god-like, pursuit.

In *Peering Down at Stones*, the poem distils the day of the field trip and a transformation of the speaker. In the opening, Hazel sets off optimistically along the beach "Seaweed squelching under careful wellies". As she walks she reminisces about her childhood experiences at the seaside, describing her movement across the beach and her anticipation for her scientific work "Who will show their faces for counting?" But in the fourth stanza the mood shifts:

> The scene is set and the pools a plenty
> Closing in on the weary subjects
> A giant to their eyes peering upward
> Glaring over my chosen metropolis

The speaker starts to consider the non-human targets of her study and expresses empathy for them, they are "weary subjects" who peer upwards in a tremulous manner, while she is a "giant" who is "glaring" powerfully over the area she has to examine, the non-human "metropolis" that she has "chosen". She then provides details in relation to the use of the quadrat:

> Eagerly placing the quadrat on the ground
> Fashioning mesmerising ripples of tiny disturbances
> Movements fast and slow of little species
> I wonder, the life of a crusty limpet

Like Daniel in *Shore-footedness* the use of the quadrat has encouraged in her a detailed observation of the non-human, and seemingly provoked an imaginative curiosity: "I wonder" as the disturbed "little species" respond to the "tiny disturbances" that she is making. She follows, observing that "I play god", moving their world around them and then embarks on her counting and data collection. The final phase of the poem depicts the end of the day, where the speaker departs, but now "tiptoeing" over the seaweed, as if treating it with a newfound respect after the day's scientific work. In her final lines there is a change of state from the opening of the poem: where before the speaker squelched energetically over the sea shore, now she regards the non-human in a different way, Clark's (2019) poetic space providing some time for reflection over the experience of doing some scientific work and reworking that in the imaginative space:

> The beach knows little of my presence
> Though I stand and thank it for its time

Threaded through these poems by the pre-service teachers is, certainly, an initial distrust or wariness by the arts graduates of being involved in scientific work.

Reflections on the project

In this chapter we have asked ourselves what part poetry might play in the language arts' contribution to environmental education. The project required that English and science pre-service teachers collaborated in scientific fieldwork, which took the pre-service teachers of English, particularly, out of their comfort zone. Using this fieldwork as a source for poetry for young people provided the pre-service teachers with novel approaches to thinking about the environment and led to a variety of responses – poems taking on the persona of the sea, the shore or the inhabitants of the shore. The pre-service teachers experimented with representing a non-human perspective and with crafting poems for a young teenage audience, with varying degrees of success. What was definitely a success was the element of collaboration between arts and science graduates, working together in the environment on nature-based tasks that informed their thinking about their teaching. Daniel, a pre-service teacher of English, in an interview after the project commented:

> Being at the seaside itself … we wrote poetry from that and there was no way we could have done that from imagining it, well you can kind of grab some stuff from your imagination from in the classroom, but actually being there and capturing some of the sea life that we found and shore life that we found and then coming back to the classroom and writing it from there, we had to be out on the trip to be able to gather those thoughts, take the photos and learn a bit as well from the scientists along the way, so yeah there is no way we could have written the poetry we wrote I don't think without having gone to X bay to do that.

Hazel, a pre-service teacher of English, observed that:

> I think it was really important to work cross-curricularly [sic] with another subject area. Personally, I found it really interesting to discuss with other colleagues from a different subject area how they would go about teaching. It definitely influenced the way that we wrote our poems and looked at the e-books and how we were going to teach that to students of both science and English, that was really important for me.

The science tutor for the Science cohort of pre-service teachers commented:

> – And I think it was extremely important for them [the pre-service teachers] to exchange their ideas … because English trainees had a different focus from the science trainees, in terms of what they were using the iPad for, while they were actually out in the field … I think that collaboration gave them examples of creativity that perhaps they hadn't thought about before.

One science pre-service teacher noted that:

> I think the main thing I can take away from that, is the sort of benefits of working collaboratively and being able to bounce ideas off one another and how sort of, your finished product, in this case the e-book, is a lot richer because you have been able to share ideas and sort of bring everything together … it sort of challenges your own way of thinking in a

> way that someone else comes along and tells you how they see things. And then you have to kind of evaluate that and think, *well where does that leave what I think?*

The project was part of a larger collaboration on the affordances of mobile technology in learning and teaching, but it was ideally suited to getting learners together to work in, for and about the environment. As teachers and teacher trainers, our awareness of environmental education has traversed a steep learning curve over the last few years and we are more than ever aware of an acute need to engage with issues in a positive and creative fashion, taking care not to dwell on "grief work" (May, 2022, p. 135), particularly with younger learners. The project demonstrated what the language arts can offer to work in this area, as it provided an opportunity for the pre-service teachers to write a range of poems for young people, bringing their skills as sophisticated English graduates to the project. In particular, the work specifically modelled ways to enable school students to engage with nature, in parallel with their scientific studies. In the way that Jessica imagined how a creature might feel, so Jorie Graham advocates that we need to "help it to be felt, help it to be imagined" (Graham, 2012, p. 40).

As teacher trainers, if we were to work on a project such as this again, we would now place much more emphasis on working with pre-service teachers to understand how embedded anthropocentric views are in literary works and to examine more ecocritical concepts and readings. On analysing the poems that the pre-service teachers wrote, we used Buell's (1996) definition of an environmentally orientated text (see earlier section) to guide our analysis. Utilising these principles, along with our use of established literary analysis methods, helped us look in new ways at what the pre-service teachers had written. Our discussions over applying Buell's criterion (iii) in relation to accountability to the environment being part of the ethical orientation of the text enabled us to draw clear distinctions between different writers' attitudes to nature present within the text. For example, Nicola's poem *Headland* and Tim's poem *Shore Shrank Creatures* position the speakers as observers, who simply look on passively, whilst poems such as Daniel's *Shore-footedness* and Hazel's *Peering Down at Stones* are both concerned with the ethics of the activities of the pre-service teachers on the shoreline. We feel that using these criteria has enabled us to add new ways of looking at texts over and above our familiar patterns of literary analysis and these criteria would be a very useful addition to ways of discussing and approaching textual analysis in schools and higher educational settings.

Conclusion

Having reflected on the project and what we consider it offered our pre-service teachers, we feel that the project has raised a number of questions which we are still considering.

We were very keen to find an opportunity to employ the medium of poetry in the project, which naturally encouraged the use of poetic techniques; however, we do not know how different the responses might have been had we asked the pre-service teachers to write in a different medium, for example observational nature prose or scientific analytical texts. As part of the process of crafting the poems, the pre-service teachers had a workshop with a published children's poet and former teacher to enable them to engage more fully in producing poems for a younger audience. If the project was repeated, we feel that some work with a poet exploring the work of ecopoets such as those included in Fisher-Wirth and Street's *Ecopoetry Anthology* could encourage them perhaps to write in less anthropocentric ways. Encountering extreme ecopoetry, such as Evelyn Reilly's *Styrofoam* (2009) or Jody Gladding's (2014) *Translations from Bark Beetle* (described earlier) might help challenge their assumptions about the use of language and perspective in their own poems. It might also help them break free of that "intellectual tameness" (2019, p. 59) which Clark notes as a characteristic feature of much current poetry about global warming, and encourage a move closer to Moe's *zoopoetics*. Finally – and perhaps in the light of the comment from Ursula Heise cited near the start of this chapter, most worryingly – we perceived in the English graduates' poems little sympathy for science. They wrote about it as if it were the enemy, rather than a powerful and essential collaborative tool. We would certainly recommend greater training in scientific literacy for English pre-service teachers. As Keller writes: working with poetry, ecopoets "create visions [which] can help us use our collective capabilities to make scientifically governed choices that are ethical in relation to humans and non-humans across the globe" (2017, p. 30). We want to encourage this scientifically informed, ethical vision for our pre-service teachers' work with their future students.

Acknowledgement

We dedicate this chapter to the memory of Sue Dymoke, poet, teacher and scholar – an inspiration to so many.

References

Bate, J. (2000). *The song of the earth*. London: Picador.

Brownjohn, S. (1980). *Does it have to rhyme? Teaching children to write poetry*. London: Hodder Education.

Buell, L. (1996). *The environmental imagination: Thoreau, nature writing, and the formation of American culture*. Cambridge, MA: Belknap Press of Harvard University Press.

Burden, K., & Naylor, A. (Eds.). (2020). *Transforming teacher education with mobile technologies*. London: Bloomsbury. Retrieved January 28, 2024, from https://www.bloomsbury.com/us/transforming-teacher-education-with-mobile-technologies-9781350095632/

Clark, T. (2019). *The value of ecocriticism*. Cambridge: Cambridge University Press.
Crutzen, P. J., & Stoermer, E. F. (2000). The anthropocene. *Global Change Newsletter, 41*, 17–18.
Durand, M. (2011). The elegy of ecopoetics. *Interim, 29*(1–2), 58–62.
Dymoke, S. (2003). *Drafting and assessing poetry: A guide for teachers*. London: SAGE Publications Ltd.
Dymoke, S., Lambirth, A., & Wilson, A. (Eds.). (2013). *Making poetry matter: International research on poetry pedagogy*. London: Bloomsbury.
Fisher-Wirth, A., & Gray Street, L.-G. (Eds.). (2013). *The ecopoetry anthology*. San Antonio, TX: Trinity University Press.
Gladding, J. (2014). *Translations from bark beetle*. Minneapolis, MN: Milkweed Editions.
Graham, J. (2012). An interview with Jorie Graham. *EarthLines, 1*(2), 36–41. Retrieved November 21, 2023, from https://www.joriegraham.com/system/files/Jorie+Graham+Interview_0.pdf
Head, D. (2022). *Nature prose: Writing in ecological crisis*. Oxford: Oxford University Press.
Heise, U. K. (2021). Introduction: Planet, species, justice – and the stories we tell about them. In U. K. Heise, J. Christensen, & M. Niemann (Eds.), *The Routledge companion to the environmental humanities* (pp. 1–10). London and New York, NY: Routledge Taylor & Francis Group.
Johns-Putra, A., & Sultzbach, K. (2022). Introduction. In A. D. Johns-Putra & K. Sultzbach (Eds.), *The Cambridge companion to literature and climate* (pp. 1–23). Cambridge: Cambridge University Press.
Keller, L. (2017). *Recomposing ecopoetics: North American poetry of the self-conscious anthropocene*. Charlottesville, VA: University of Virginia Press.
May, T. J. (2022). Climate theatre: Enacting possible futures. In A. D. Johns-Putra & K. Sultzbach (Eds.), *The Cambridge companion to literature and climate* (pp. 132–145). Cambridge: Cambridge University Press.
Mentz, S. (2024). *An introduction to the blue humanities*. Abingdon and New York, NY: Routledge.
Moe, A. M. (2018). Toward a poetics of Gaia: Biosemiotics and Jody Gladding's translations from bark beetle. *ISLE: Interdisciplinary Studies in Literature and Environment, 25*(3), 450–467. https://doi.org/10.1093/isle/isy054
rawlings, a. (2006). *Wide slumber for lepidopterists*. Toronto, ON: Coach House Books.
Reilly, E. (2009). *Styrofoam*. New York, NY: Roof Books.
Reilly, E. (2011). The grief of ecopoetics. *Interim, 29*(1–2).
Spahr, J. (2011). *Well then there now*. Jaffrey, NH: A Black Sparrow Book.
World Bank. (2023). *World Bank group country classifications by income level for FY24 (July 1, 2023–June 30, 2024)*. Retrieved November 21, 2023, from https://blogs.worldbank.org/opendata/new-world-bank-group-country-classifications-income-level-fy24

10 What can storytelling and drama offer to students' ecological understanding and citizenship? *Being trees*

Molly Mullen, Mark Harvey, Marie McEntee, Christina Houghton, Ariane Craig-Smith, Hannah Neville and Debbie Larkins

As Māori, our origins are in the natural world, including the trees.
(Lee-Morgan, 2019, p.154)

Introduction

This chapter shares three examples of drama and storytelling practice in Aotearoa New Zealand primary school classrooms, which set out to involve children in biosecurity issues related to forest health. In each example, drama and storytelling were interwoven with Western science and mātauranga Māori (Māori knowledge). We consider the extent to which drama might engender creativity and agency in relation to biosecurity, as children produce and act within story-worlds where they encounter (feel/embody), explore and express a sense of ecology, interspecies empathy and species interdependence. We are interested in how drama, as part of a shared inquiry, might support children into different forms of creative eco-citizenship.

What are the key issues?

For an island nation like Aotearoa New Zealand, biosecurity is essential for protecting the country's unique biodiversity, and particularly its vulnerable flora and fauna, against harmful pests and diseases. This protection occurs both at the border to prevent unwanted organisms entering, and inside the border to eliminate those pest organisms that may have already established themselves. The Aotearoa New Zealand government agencies involved in managing Aotearoa New Zealand's environment seek to involve people in caring for the environment, and biosecurity initiatives aim for Aotearoa New Zealanders to be involved as a "team of 4.7 million" (Ministry for Primary Industries, 2016), whereby people of all ages are invited to play an active role in ensuring a resilient biosecurity system. While children are viewed as target audiences and "receivers" of biosecurity messages, or as "conduits" for transferring key biosecurity messages to their whanau (family), they are not typically included and engaged in biosecurity conversations, and do not have consistent access to

DOI: 10.4324/9781032634524-14

knowledge about ngahere ora (forest health) (Ram, France & Birdsall, 2016). To address this gap in biosecurity engagement, we created Toitū te Ngahere (TTN), a project exploring how mātauranga Māori, creative arts and the social and environmental sciences in schools might support children's, teachers' and their community's awareness of, and responses to, two biosecurity threats, kauri dieback and myrtle rust, which are having devastating effects in Aotearoa New Zealand's native forests.

Aotearoa New Zealand is a country in which forests form a significant part of the natural landscape. Before the arrival of people, dense native forest and shrubland covered more than 80% of the land (Ministry for Primary Industries, n.d.). Clearance for agriculture, timber and settlement, beginning with the arrival of Māori ancestors (approximately 800–1000 years ago), and greatly increasing as settlers began to arrive from Europe in the early 1800s, has reduced this to 38% forest cover, of which approximately 4/5 is native forest, with the remaining 1/5 exotic and plantation forest (Ministry for Primary Industries, n.d.). This rapid clearance, along with other environmental pressures, has contributed to a situation in which 46% of native vascular plants, and a great many more fauna species, are threatened or at risk of extinction (Stats NZ, 2023).

Aotearoa New Zealand is home to many endemic species and unique habitats, which evolved in long isolation from other land masses. Developed within this unique environment, the Indigenous Māori culture encompasses unique mātauranga (knowledge), embedded in stories and cultural practices intimately related to different native species, meaning native trees are often interwoven with aspects of cultural identity. For other Aotearoa New Zealanders, native forests and specific tree species also have deep emotional, social and economic importance. The identification of new plant diseases that impact native species, therefore, has complex ecological, cultural, social and economic implications.

Two diseases impacting taonga (treasured) tree species are having particularly devastating impacts. Kauri dieback disease (*Phytophthora agathidicida*) was found in 1972 and finally identified in 2008. It has since been found in seven key remnant kauri forests in the North Island of Aotearoa New Zealand. Kauri (*Agathis australis*) are ecosystem engineers, which means kauri dieback disease, which is deadly to these trees, poses risks to co-evolved dependent native species (Bradshaw, Bellgard, Black, Burns, Gerth, McDougal ... & Horner, 2020). Myrtle rust (*Austropuccinia psidii*) is an airborne fungal disease that impacts species within the myrtaceae family of plants, of which Aotearoa New Zealand has 27 identified endemic species. It arrived in Aotearoa New Zealand in 2017 and its rapid spread has already devastated some native myrtaceae and threatened others (Toome-Heller, Ho, Ganley, Elliott, Quinn, Pearson & Alexander, 2020). Community participation and engagement have been a significant part of responses to these threats (Lambert, Waipara, Black, Mark-Shadbolt & Wood, 2018; Hill, Ashby, Waipara, Taua-Gordon, Gordon, Hjelm, Bellgard, Bodley & Jesson, 2022).

For some time now, researchers and educators have argued that the arts have multiple important roles to play in environmental education (cf. Bergmann, 2010). There is now an expansive body of literature showing how experiencing, interpreting, making and/or sharing artworks can all impact students' knowledge about and feelings towards the natural world. In addition to these outcomes, TTN wanted children to be recognised (by themselves and others) as contributing meaningfully to biosecurity and ecological care in their communities (and beyond). In designing TTN, therefore, we drew on a previous project from 2017, when one member of the TTN team had worked with three schools, using knowledge and methods from coastal geography, social science and science communication, to take action in response to ocean pollution. We were also informed by Mary-Ann Hunter and her colleagues in Tasmania, who found school-based, artistic inquiry projects offered children opportunities to connect with each other, communities and the "material world" and, through making and sharing their artworks, to be acknowledged "as capable, useful citizens" (Hunter, Aprill, Hill & Emery, 2018, p. 98).

Toitū te Ngahere (TNT) involves Māori and Pākehā (NZ European) university researchers working in partnership with schools and their communities to learn about forest health from each other, and from others where needed. It aims to go beyond informing children about the two diseases and ensuring they understand the official biosecurity responses. Instead, with each school, the research team and teachers co-plan a cross-curricular inquiry process lasting approximately two terms (20 weeks). TTN team members then work alongside each class, offering ideas and looking for ways to support the overall inquiry aims, as well as questions and ideas emerging along the way, suggesting resources, bringing in outside experts or running workshops. The inquiry processes include a mix of activities led by the class teachers: student-led activities; workshops with researchers, talks or activities with visiting experts and knowledge holders, and school trips. The project methodology moves flexibly through the following phases:

Whanaungatanga: Connections and relationships formed and consolidated, through family building processes.

Pūrākau: Stories of origin, that are meaningful to each iwi/hapū (tribe/subtribe).

Wānanga – Whakamārama: A process of teaching, learning and exchanging towards illuminating understandings (in this case of ngahere ora (forest health) and ecology).

Toi Wānanga – Whakamārama: As above, only through art forms explored.

Toi Whakapiri: Children create artworks related to forest health in ways that are relevant to their whakapapa (genealogies), place/environment and community.

Whakamana: Public Art installed (whakamana is used here to mean uplifting the mana, or integrity, charisma, well-being, spiritual state and authority amongst other things).
Whakatakotoranga: Presentations to audiences (from new learnings).
Whakaaroaro te mahi: Reflecting on the presentations and work done in the project.

The project has run for two and a half years, involving over 300 children and teachers from five schools in Tāmaki Makaurau, Auckland. Schools have worked across multiple art forms over the duration of the project. In this chapter, we focus on their storytelling and drama work.

What does the research literature say about the role of storytelling and drama to foster ecological understanding and eco-citizenship?

In European contexts, stories are recognised as important to how people encounter, express and share complex knowledge about ecology and other species and foster a sense of kinship and interdependence (Nanson, 2021; Brown, 2013). But, as Anthony Nanson (2021) points out, because the terms story and storytelling are applied to vastly different media and practices, it is important to be specific. In this chapter we discuss two forms of storytelling relevant to TTN: the Māori forms and practices of pūrākau and pakiwaitara, and secondly the creation or creative retelling of stories about forest health using drama.

Pūrākau and Pakiwaitara

One of the key Māori cultural practices TTN has drawn on with the school children is the telling of pūrākau. Throughout te ao Māori (the Māori world) pūrākau are well-known to be stories of origin (Cliffe-Tautari, 2020). As Jenny Lee-Morgan (2019) states, "*Pūrākau* literally refers to the *pū* (base or foundation) of a *rākau* (tree)" (p. 151). Citing Lee (2009), Dan Hikuroa (2016) explains, "Pūrākau [are] a traditional form of Māori narrative, containing philosophical thought, epistemological constructs, cultural codes, and worldviews that are fundamental to our identity as Māori" (p.6). Pūrākau have significant whakapapa, being widely regarded in Māori contexts as dating back at least three to four generations. Some can be traced back to early Māori ancestors. Lee-Morgan provides a summary of common pūrākau across te ao Māori (the Māori world), emphasising human links and origins in connection with trees as follows:

> One of the most famous children of Papatūānuku (Earth Mother) and Ranginui (Sky Father) is Tāne Mahuta (God of the Forest). There are numerous *pūrākau* about Tāne's memorable feats, including the story of separating his parents from their tight embrace so there would be

> space and light for their descendants to develop – the trees, the shrubs, the ferns, and eventually all living things. In this particular *pūrākau* he takes the form of an enormous tree that grows with power and strength to push his parents apart. It was also Tāne Mahuta who created the first woman, Hineahuone, with whom he begat the first child. There are multiple *pūrākau* about Tāne; epistemological designs that provide a deep insight into the ontological constructs of our world, our wisdom, and our potential. At its simplest level, the *pūrākau* of Tāne Mahuta tells us that we are the trees, and the trees are us.
>
> (2019, pp. 154–155)

Pūrākau are often iwi/hapū (tribe/sub-tribe) specific. For example, different iwi/hapū are likely to have different pūrākau about how tohorā (southern right whales) are connected to the whenua (land). One such pūrākau was shared with the children taking part in TTN in 2023 by kaumātua and iwi representative Rodney Ngāwaka (Ngāti Rehua, Ngātiwai iwi): how his hapū of Aotea (Gt. Barrier Island) have pūrākau about how tohorā come to their island and rub against the outlying rocks. Another one described how Aotea is the leftovers of the ika (fish) that Māui fished up (with for instance its mountains being the spine of the fish).

Pūrākau have historically often been misunderstood and twisted away from their earlier intentions, particularly outside Māori contexts and because of colonisation. Deliberate attempts have been made to replace mātauranga Māori with European knowledge frameworks like Christianity and Western science (Lee, 2009). Within these frameworks, pūrākau "have historically been misappropriated, misrepresented, and misinterpreted as 'fables' or 'anecdotes'" (Cliffe-Tautari, 2020, p. 13), not to be taken seriously. However, as Tania Cliffe-Tautari (2020) notes, "pūrākau remain critical to Māori, as they preserve the cultural repositories of past generations. Not only that, pūrākau offer Māori huge pedagogical potential, as they allow us to transcend time and space" (2020, p. 13). Lee (2009) notes that engaging with pūrākau is one way in which Māori engage with the world as a complex way of being.

Pūrākau are shared for multiple reasons, such as in reclaiming stolen land through court processes, and when decolonising educational contexts from Māori perspectives (Lee, 2009). Lee proposes that pūrākau can even be used as a Māori research method towards cultural and political empowerment. TTN has engaged with pūrākau as one way of reclaiming Māori space, and as a culturally responsive, place-based way to inform and empower children to learn, engage with and reflect on ecological issues. This aligns with what Penetito (2009) describes as "*ecological consciousness*", a deeply interconnected sense of who you are, and how you fit into a place/environment (p. 19). For example, in 2022, Pita Turei (Ngai Tai Ki Tamaki, Ngāti Pāoa, Ngā Rauru Kiitahi iwi), storyteller and iwi advocate, visited each of the two schools on mainland Tāmaki Makaurau (Auckland), each time sharing pūrākau that connected to and expressed knowledge about the land on which the schools were located.

Turei also used a specific pūrākau to invite students to think about their role in supporting mana whenua (Māori iwi/hapū with designated authority over land/waters), and their care for land and forests.

Under Māori guidance, and with Māori and Pākehā co-leadership in our team, we have used pūrākau to guide students to generate stories that are in conversation with and reflect on kauri dieback and myrtle rust. The students have been supported to tell these stories in different mediums, including live theatre, video and animation. Rather than call the stories the students have made pūrākau and misrepresent what they are, we are calling them pakiwaitara (stories or narratives). Some of these pakiwaitara appear as unique and contemporary yarns, while others attempt to extend on pūrākau with imaginary "what happens next" scenarios.

Making and telling stories through drama

There is broad agreement, across literature on drama education and environmental issues or ecology, that involving students in embodied, collaborative creation through drama can enhance children's understanding of complex ecological and environmental knowledge (Raphael, 2021; Davis & Tarrant, 2014; McNaughton, 2010, 2006, 2004). Research by science and environmental education scholars using drama comes to similar conclusions (Bailey & Watson, 1998; Çokadar & Yilmaz, 2010; Curtis, Howden, Curtis, McColm, Scrine, Blomfield, Reeve & Ryan, 2013). For example, when compared to more traditional classroom methods for teaching ecological concepts, the active, collaborative, imaginative and affective nature of drama-based pedagogies can lead to enhanced understanding (Çokadar & Yilmaz, 2010). It also seems that different forms of drama and different kinds of participation can affect student understanding. Curtis et al. (2013) found students' knowledge about climate change was most enhanced when they devised their own plays in response to climate change science rather than watching plays about the topic. Students who collaboratively created their own plays also became more confident in sharing their knowledge. Devising drama involves multiple learning processes: investigating, exploring, interpreting and representing knowledge through movement, speech and sound, symbol, spatial and physical relationships, as well as through research, discussion and writing. As such, it can be understood as a process of inquiry that allows for the consolidation of prior learning but also for new observations, insight and understanding to emerge. Curtis et al. (2013) suggest that opportunities for students to devise plays or create dramatic worlds allow students to engage in a deep, multi-layered exploration of environmental issues.

In addition to enhancing understanding of the environment and ecology, Marie McNaughton (2004, 2006, 2010) proposes that drama prepares students to take informed environmental action. Through drama McNaughton's child participants, from Scottish schools, were able to step inside local environmental issues and apply their knowledge to formulate and try out possible solutions or alternatives within the fictional story. McNaughton (2006) suggests drama enables children to "rehearse active citizenship" (p. 139); as children take roles within a fictional world they become "active participants in the events, tensions, problems and solutions" of the story (p. 154). In doing so, children gain skills in, for example, co-operative problem solving, which can be applied to "real" environmental scenarios in the future. This idea expresses a broadly "action competence" model of citizenship, which conceives of action as following on from the right knowledge, attitudes and skills and of humans as the primary agents of sustainability. More recent drama education research, however, reflects shifts in thinking about children's citizenship and eco-citizenship, towards more child- and ecology-centred models (Hickey-Moody, Knight & Florence, 2021; Hatton, 2022; Heggen, Sageidet, Goga, Grindheim, Bergan, Krempig, Utsi & Lynngård, 2019).

Much contemporary thinking about child citizenship views children as already legitimate citizens, participating in communities and "publics" in their own ways (Hickey-Moody et al., 2021). Anna Hickey-Moody brings together this conception of child citizenship with posthuman thinking to consider how children participate alongside animals, elements, plants to enact care for their environment. The arts have an important role to play, as Hickey-Moody explains: "multimodal forms of expression through art and sensuous mapping allow children to articulate the complexities of their posthuman civics, and call publics to attention in response to the anthropocene" (2021, p. 19). Marianne Presthus Heggen and colleagues (2019) also challenge the idea that children can only practise for future citizenship. They view children as already being and becoming citizens as they "participate, inhabit, negotiate and challenge the established community", including through imaginative play (p. 389). Like Hickey-Moody, they propose that eco-citizenship involves participation in ecological systems along with animals, nature, elements, weather and so on, who are recognised as part of the established community. This conception of eco-citizenship seems consistent with proposals made in recent literature on drama, story and performance, which emphasises the importance of attentive, place-based engagement with the materiality of the natural world and of fostering interspecies connections and empathy (Aaltonen, 2015; Hatton, 2022; Niemelä, Laulumaa, Tupala & Raatikainen, 2023). Aaltonen (2015), for example, describes a participatory performance project with children in a Norwegian forest which aimed to invite children to listen to and empathise with trees. Through the performance, children encountered characters that embodied knowledge about trees from botany and folklore and were invited to experience the forest through all their senses. Aaltonen proposes that the children experienced an expanded, ecological sense of identity and community, and enhanced empathy for the trees. Aaltonen shows how performances could be created for children in ways that open opportunities for eco-citizenship. In this chapter, however, we are interested in children creating and sharing their own performances. In the rest of this chapter, we consider how eco-citizenship can be enacted when children create and play within stories in ways that involve close attention to place and ecosystems, involve multi-species role-taking and make space for multiple knowledges.

Case studies of successful practice

Konini Primary School

Konini Primary School is in Glen Eden, Tāmaki Makaurau (Auckland). It has a strong environmental kaupapa (approach/platform/purpose), developed over recent years, and is a Green-Gold Enviroschool[1] TTN worked with three year-five classes at Konini, about 90 children aged nine and ten. We collaborated with their three teachers to plan an inquiry into kauri dieback and myrtle rust, which aligned with their existing inquiry topic for the term. Each class would work towards presenting art pieces and other classwork at the school's Matariki open evening, celebrating the Māori New Year with children, families and friends. The teachers worked with the children in each class to choose a main artform. One chose photography, the second chose sound art and the third chose drama, which is the class we focus on.

The drama inquiry project focused on myrtle rust and the children started to look for information about the pathogen online. We had already talked about how there was less public awareness of myrtle rust (than kauri dieback, for example), and the class noticed a lack of resources aimed at children. They decided to create dramatic scenes showing how myrtle rust spread, and what humans were doing, or not doing, to manage that spread. The class teacher regularly used drama in her class and felt confident with the devising process. TTN team members worked alongside her to run a small number of drama, movement and costume-making workshops.

Early in the process, following a recap of what the children knew about myrtle species and myrtle rust, we introduced games which invited the children to move and play to create a forest ecosystem. The aim was to create a felt sense of how the life forms and elements in a forest are interconnected, so when one is affected so are the others. In the next workshop, the children chose a tree, plant or animal species to embody. We used creative drama and movement tasks to invite the children to experiment with movements and sounds for their creatures. The children split into three groups, each with a scenario of myrtle rust spread to work with. Group 1 created a scene of a person buying a plant from a nursery and taking it home. Group 2 showed myrtle rust spores being transferred from plant to plant by birds and bees. Group 3 showed photographers visiting a tree and then returning later to see it covered in rust. They initially improvised the scenes using movement and sound. They then refined the structure and added some dialogue or narration. The children went through multiple stages of making, presenting and giving each other feedback. The teacher also led reflective discussions, encouraging the students to connect their scenes to their wider knowledge and local contexts. For example, after reflecting on relevant mātauranga Māori, some groups introduced characters and concepts from te ao Māori into their scenes.

Figure 10.1 Screenshot, "Tāwhirimātea blows the myrtle rust onto the fighting animals" from Wind Blown Parasite, a video created by Konini School students, 2022

One creative challenge was to visually show the spread of the microscopic rust spores. At a costume-making workshop, the children tried out different ways to represent the bright yellow spores and their spread by wind. They settled on using stuffed yellow balls, the size of a tennis ball. These were attached to one plant or animal costume using sticky velcro squares and could then be detached and moved to a new host by the children as puppeteers (as if by the wind for example). The scenes were made into short videos, with key text placed between each episode of action.

> *"The myrtle Rust has murdered the Kānuka" … "The Myrtle Rust has struck again!"*
> *"Thank goodness the Myrtle beast is gone. What are those yellow spots on the Ramarama"*

[Link to this video: https://www.mobilisingforaction.nz/ra1-4-teuru/#video]

The children at Konini put their knowledge of how myrtle rust spreads and affects the forest into action by working playfully with their bodies, sounds and costumes. Their scenes revealed an ecologically attuned, dramatic story of who lives and who dies in their multi-species community. The dramas conveyed the implications of an invasive pathogen in complex systems of interrelations between animal and plant species, people and human practices. Within their own vernacular, the children expressed empathetic care for the creatures of the forest. The titles of the dramas created by the children, Wind Blown Parasite, The Killer Powder, The Cruel Killer & Murder Rust, expressed how the children felt about what they had learned. They recognised and wanted to represent the horrific reality of what is happening in their forests as a call for action. To do this they drew on the vernacular of scary movies, crime stories and action heroes, portraying the dramatic death of the myrtle species and using urgent, comic-book-like dialogue and titles: "Murdered by myrtle rust", "myrtle rust strikes again".

In interviews conducted before the dramas were created, many Konini students described their close relationship with the native bush and showed they felt the seriousness of the myrtle rust threat deeply. One girl said, "I'm scared I don't wanna go in [to the forest] 'cause I'm scared I'll spread something" (Konini Pupil Focus Group 1, 2022). We were aware of research indicating the importance of approaching environmental issues with children in ways that create conditions for hope rather than fear, but also that this does not mean expelling all feelings from classrooms (Raphael, 2021). Raphael suggests: "feeling the threat matters can be the very thing that causes us to want to do something about it. Experiencing and exploring these emotions together with others, through the protective frames that drama allows, can be reassuring" (p. 10). Raphael echoes arguments from Bolton (1984), and Dunn, Bundy and Stinson (2015) "that our responsibility as facilitators is not to protect our participants from emotions but rather, into them" (p. 5). Of course, not all students in a class will have identical emotional experiences, and students' feelings can change moment-to-moment in response to complex factors within and beyond the drama (Dunn, Bundy & Stinson, 2015). But, as Raphael argues, when working with environmental topics it is important to acknowledge the potential for students to experience eco-anxiety. We sought appropriate ways to respond to the feelings expressed by the children at Konini, both within and beyond the drama work, to facilitate possibilities for hope and action (Raphael, 2021). Consistent with the ideas presented by Raphael, in the process of making their scenes the Konini students shared their worries with each other, their teachers and the wider community, and, as indicated by subsequent interviews, felt that their feelings were recognised and validated. The students were able to experience how their fears for the forest they cared about so deeply could be channelled into the creativity of producing their scenes. In addition, the sharing of the scenes enabled the children to see their messages being received with due consideration for the concerns they had.

The video-recorded scenes were played at a Matariki exhibition at the school and later installed in a local gallery, alongside the artworks of the other children. They were also shown by the children in a lecture for first-year university students and put online. In a reflective interview, we asked the children whether they thought showing their videos made a difference. One boy responded:

> it does, it does a lot ... teaching people that it's not something like COVID that could pass on, it's something that maybe will stay forever. So that's the scariest thing. Because what if it just stays, the disease stays living for that long? I don't think a lot of people would know a lot about myrtle rust and we've learned so much that we can teach a lot of people. So Te Uru is going to be amazing for people to recognise that it's so just so scary.
>
> (Konini Pupil Focus Group 2, 2022)

By entering physically and emotionally into the ecology of the forest through play and drama, we suggest that the children of Konini created informed, emotionally powerful scenes, which were both humorous and deadly serious, acting on their felt sense of fear and urgency to try and instil a similar impetus for action in their audiences.

Te Kura o Okiwi

Te Kura o Okiwi (The School of Okiwi) is located at the north end of Aotea Great Barrier Island, which is located 90 km off the east coast of Tāmaki Makaurau Auckland. It is one of three full primary schools on the island, each with roughly 30–40 students from years zero to eight, ages five to 13. Okiwi is the only Māori medium school on the island, where all students are taught curriculum subjects in te reo Māori for over half the time. The kura is split into two groups, *tuākana* (elder brothers and sisters) and *tēina* (younger brothers and sisters), who have separate learning spaces but work together on many activities and projects. Okiwi has an established, deep engagement with te taiao (the environment) and mātauranga Māori. The children are active in caring for the school environment and the wider environment of the island. They work closely with Mana Whenua, Ngāti Rehua Ngātiwai ki Aotea and put te ao Māori at the centre of their curriculum. Whilst the artistic choices in TTN are primarily made by the students, it was the kaiako (teachers) at Okiwi who initially suggested drama, as they were keen to provide space for their students to engage their bodies, voices and imaginations, and wanted to build the confidence of the older children. The school also wanted mātauranga Māori, and the kaiako suggested exploring a pūrākau that connected to TTN's themes "Te Tohorā me te Kauri" (The Whale and the Kauri).[2]

Our first workshop introduced drama using games and then invited the children to step into and explore the world of the pūrākau. First, we worked with all the children, teina and tuākana, together. They began the session with a karakia (Māori prayer). We introduced drama as a form of play and storytelling through a series of games that involved exploring their bodies, space, objects and relationships in imaginative and creative ways. This involved working together as a whole group and in small teams. The games generated energy, lots of laughter and a sense of appreciation for what each other shared. We then worked separately with the tēina, who returned after their break asking to do more drama. We introduced them to the characters of the whale and the kauri tree, who are siblings in the pūrākau. In two groups they worked out how to show each character using their bodies as a group. They took considerable care in representing each character accurately. The kauri group carefully considered each layer of the forest ecology, an idea introduced in an earlier TTN workshop. We then read a version of the pūrākau once through. For the second reading, we used a drama strategy known as a "Whoosh", re-reading the story so that every child can participate and take multiple roles over the duration of the telling. Their enthusiastic and imaginative physical enactments of the trees, forest creatures, whales and ocean delighted everyone involved. Finally, we asked which parts of the story they wanted to experience again. They chose four key moments from the pūrākau, and we helped them distil these into four tableaux or still images before discussing what made those moments important. These moments became the key scenes in their final play. In the next workshop we helped the class work out how to tell each of the four moments of the story through movement, considering how to show the characters' relationships, feelings and desires, and a moment of conflict or change in each section. The teacher then rehearsed and further developed the scenes, helping the children to write dialogue and narration, and to create waiata (songs) for the transitions between each scene. The children also had the idea of adding a new section to the story, where the kauri had got sick, and the whale came back to offer itself to help heal the tree. This drew on the children's knowledge of a mātauranga Māori research project using whale bones to create rongoā (Māori medicine/treatment) for kauri dieback. From the first workshop, this group always chose to have Tāne Mahuta present on stage. The children and kaiako came up with initial ideas for costumes and scenery, which we helped them to make on a third visit. The children performed their play for the rest of the school and their whānau (family).

With the tuākana, we used drama to explore the theme of connection and disconnection from the pūrākau, they were interested in the story and the way a contemporary retelling connected it to forest health issues. But they were not all enthusiastic about the idea of performing. On our second visit, then, we gave them different options for the forms they could use, and they settled on stop-frame animation. This group eventually created a series of short animations showing key moments from the pūrākau as well as imagined scenes that came afterwards, showing how people became disconnected from the ngahere (forest) undertaking actions that had damaged its health, and then how people today were working together to restore the well-being of the ngahere. The videos were shown to the tēina and whānau and displayed for the wider community at the island's Ecology Festival. Their videos conveyed a sense of the ecological and cultural implications of the devastation caused by threats like kauri dieback and myrtle rust and of the different ways humans have related to and valued the natural world.

Figure 10.2 Students at Te Kura o Okiwi rehearsing their drama onsite at their school

Becoming a tree can be seen as a cliché in drama classrooms and is avoided altogether by some in our experience. The tēina at Okiwi, however, never seemed to stop finding delight in embodying whales, kauri trees and other elements in the story. Even in a final interview, more than six weeks after their performance, the children spontaneously leapt up to show their favourite moments from their play, rather than describing them. Their portrayal of the creatures showed detailed knowledge of them and respect for them, the quality of the bark of the tree or the skin of the whale, the importance of the root system. But the children's multi-species role-taking can be understood to be about more than expressing knowledge. Jo Raphael (2021), writes:

> Playing roles of human and non-human elements within the environment can help shift participants from an anthropocentric view of the world as they view it from the point of view of other living things. Drama can build empathy, understanding of the interdependence of ecosystems, and a sense of connection to all life on Earth.
>
> (Raphael, 2021, p. 9)

Further, writing about pūrākau and Indigenous storywork, Jenny Lee-Morgan's quote from earlier in the chapter (Lee-Morgan, 2019) about trees and Māori reinforces that in te ao Māori we are trees and trees are us. Becoming trees and whales in the classroom might be one way of exploring identity and culture as interconnected to te taiao (the environment), connected for Māori directly through whakapapa. Raphael (2021) suggests that playing non-human roles in drama can shift anthropocentric perspectives. In the context of Okiwi, however, it might be better understood as one way, pedagogically, to sustain a non-anthropocentric ontology (Lee-Morgan, 2019).

Kauri Park School

In this final example, we consider how children's experience of drama and performance might inform subsequent projects and the children's ongoing desire to be protectors of and advocates for forest health. Kauri Park School is based in Beach Haven on the north shore of Tāmaki Makaurau's Waitemata harbour. Kauri Park is an Enviroschool, with multiple environmental projects involving children, staff, families and the community. The school is also well-resourced for the arts, with its own radio station. TTN worked with five classes of mixed year five and six pupils, aged nine to eleven. In the first year, the teachers at Kauri Park decided to incorporate the theme of forest health into a large-scale musical production. The performance involved all 120 children, and was performed for the rest of the school, families and the local kindergarten. The play was also a celebration of Matariki, with each scene exploring mātauranga Māori about each star in the Matariki cluster. In the play, a group of children were transported through different habitats, meeting each star, as well as the animals, plant species and other aspects of te taiao protected by the atua (gods).

The TTN team's conversations with students in focus groups revealed how much the children from Kauri Park valued their role as "teachers" of ngahere ora (forest health), as the following quotes illustrate:

> I think I'd tell [people] how [kauri dieback] spreads to give them understanding about why we clean our shoes and why we should do it, or how myrtle rust flies through the air.
>
> To tell people that this [kauri dieback and myrtle rust] is happening and that you should be aware of it … and that you can help as well.
>
> We want to bring on generations and make sure the plants are [still there] … like when you have kids.
>
> (Kauri Park Focus Group 1, 2022)

The performance provided a creative channel to outwardly express these desires and feelings through storytelling. The drama also allowed the children to express their sense of being guardians of the forest, for example in the heartfelt singing of the closing song, *Guardians of Papatūānuku*, by Siu Williams-Lemi (2021).

Figure 10.3 Mural in progress at Kauri Park School

The performance catalysed an unplanned follow-on project the following year with a group of year six children, who had been in the performance. This new project initially focussed on the children designing art posters which could be displayed on the gates of a local park that had been closed by the Council to prevent the spread of the forest pathogen kauri dieback. On the posters the children would tell their stories about forest health to raise community awareness about why the local park had been closed to the public. However, the project quickly blossomed into a much larger endeavour, when the canvas for the artwork expanded from small posters to two rusty old shipping containers that were stored on the school grounds. The containers would provide a "hard" canvas on which the children could collaboratively design and create artwork about forest health for their school community and the hundred or more households that looked across the valley at the containers. Furthermore, by scanning a large QR code on one side of the containers, the community could use their phones to directly find out about this child-led inquiry into ngahere ora, and at the same time learn about the forest pathogens kauri dieback and myrtle rust.

Under the guidance of an artist and three of the TTN team, and over the course of 20 weeks, the children developed the stories and themes from their performance into collective artwork for the shipping containers. The project became an exploration into how art making, including graphic drawing and graffiti art, can be used to communicate children's learning about ngahere ora. The stories that had been created in the drama some months earlier had now found a "permanent" home on the shipping containers that could be viewed from outside the school grounds. The children were now spreading their messages about how to be guardians of the forest within and beyond the school gates.

Closing remarks and recommendations for future practice

Toitū Te Ngahere (TTN) is a project exploring the potential of art in schools for forest health. The aim is to interweave art with mātauranga Māori and science (including social science) to support children to inform, develop and tell stories about their own relationship to the ngahere, and to take action in different ways in response to biosecurity issues. Like Hunter and colleagues (2018), we suggest that the interweaving of the arts with other knowledges about the environment can lead to creative, research-informed processes in which children are viewed to be "capable, useful [eco]-citizens" in the here and now (p. 98). In this chapter, we have also drawn on Raphael (2021) to highlight the ways drama teaching might acknowledge and respond to children's deeply felt care and concerns for the ngahere. Reflecting on their research, Hunter et al. caution that such outcomes are contingent on schools and teachers having a holistic understanding of the arts, expansive views of sustainability and openness to emergent practice. In previous writing on TTN, we also propose that, where mātauranga Māori is being interwoven, a degree of oho (to be awake) to that knowledge is also essential (McEntee, Harvey, Mullen, Houghton & Craig-Smith, 2023). A high degree of care and guidance by cultural experts is important when exploring Indigenous concepts and terms, as even common concepts can be misinterpreted (see McEntee et al., 2023). TTN's partner schools, teachers and communities have been confident in, or open to, undertaking emergent, child-centred enquiries that use the arts as a mode of inquiry and interweave multiple knowledges. The strong leadership and teaching cultures of each school have been a critical factor. We propose that this kind of approach also depends on wider systemic factors, like a curriculum that allows for inquiry-based learning and has space for Indigenous knowledge and artistic practices.

Informed consent

The project has approval from the University of Auckland Human participant ethics committee, which includes consent to name the schools and use images where consent has been given.

Acknowledgements

The TTN team acknowledges and thanks all our partner schools for their collaboration and support on this project.

Notes

1 Enviroschools is an Aotearoa New Zealand-wide programme for early childhood centres and schools that commits to long-term sustainability and environmental action. There are three stages to the Enviroschool programme. Once a school has completed all three, it becomes a Green-Gold Enviroschool. More Information can be found here: https://enviroschools.org.nz/about-us/

2 See here for the pūrākau of the kauri and the whale as told by Ngāpuhi and Ngāti Wai iwi: https://tehiku.nz/te-hiku-tv/haukainga/8211/te-tohora-me-te-kauri).

References

Aaltonen, H. (2015). Voice of the forest: Post-humanism and applied theatre practice. *Research in Drama Education: The Journal of Applied Theatre and Performance, 20*(3), 417–421. https:// 10.1080/13569783.2015.1059263

Bailey, S., & Watson, R. (1998). Establishing basic ecological understanding in younger pupils: A pilot evaluation of a strategy based on drama/role play. *International Journal of Science Education*, *20*(2), 139–152. https://10.1080/0950069980200202

Bergmann, I. (2010). What is the role of the arts in the face of this?: An exploration in the context of systems thinking and the transition to sustainability. *PAN: Philosophy Activism Nature, 7*, 23–32.

Bolton, G. (1984). *Drama as education: An argument for placing drama at the centre of the curriculum.* Harlow: Longman.

Bradshaw, R. E., Bellgard, S. E., Black, A., Burns, B. R., Gerth, M. L., McDougal, R. L., … Horner, I. J. (2020). Phytophthora agathidicida: Research progress, cultural perspectives and knowledge gaps in the control and management of kauri dieback in New Zealand. *Plant Pathology*, (69), 3–16. http://dx.doi.org/10.1111/ppa.13104

Brown, L. (2013). Storytelling and ecological management: Understanding kinship and complexity. *The Journal of Sustainability Education*, *4*, 1–6.

Cliffe-Tautari, T. (2020). Using pūrākau as a pedagogical strategy to explore Māori cultural identities. *Set*, *1*, 12–18. https://doi.org/10.18296/set.0156

Çokadar, H., & Yilmaz, G. C. (2010). Teaching ecosystems and matter cycles with creative drama activities. *Journal of Science Education and Technology*, *19*(1), 80–89. https://doi.org/10.1007/s10956-009-9181-3

Curtis, D. J., Howden, M., Curtis, F., McColm, I., Scrine, J., Blomfield, T., Reeve, I., & Ryan, T. (2013). Drama and environment: Joining forces to engage children and young people in environmental education. *Australian Journal of Environmental Education, 29*(2), 182–201. https://doi.org/10.1017/aee.2014.5

Davis S., & Tarrant, M. (2014). Environmentalism, stories and science: Exploring applied theatre processes for sustainability education. *Research in Drama Education: The Journal of Applied Theatre and Performance, 19*(2), 190–194. http://dx.doi.org/10.1080/13569783.2014.895613

Dunn, J., Bundy, P., & Stinson, M. (2015). Connection and commitment: Exploring the generation and experience of emotion in a participatory drama. *International Journal of Education & the Arts*, *16*(6), 1–21.

Hatton, C. (2022). Drama as a pedagogy of connection: Using Heathcote's rolling role system to activate the ethical imagination. In M. McAvoy & P. O'Connor (Eds.), *The Routledge companion to drama in education* (pp. 153–165). Abingdon and New York, NY: Routledge.

Heggen, M. P., Sageidet, B. M., Goga, N., Grindheim, L. T., Bergan, V., Krempig, I. W., Utsi, T. A., & Lynngård, A. M. (2019). Children as eco-citizens? *Nordic Studies in Science Education, 15*(4), 387–402. https://15.10.5617/nordina.6186

Hickey-Moody, A., Knight, L., & Florence, E. (2021). *Childhood, citizenship, and the anthropocene: Posthuman publics and civics.* Lanham, MD: Rowman & Littlefield.

Hikuroa, D. (2016). Mātauranga Māori—the ūkaipō of knowledge in New Zealand. *Journal of the Royal Society of New Zealand*, *47*(1), 5–10. https://doi.org/10.1080/03036758.2016.1252407

Hill, L., Ashby, E., Waipara, N., Taua-Gordon, R., Gordon, A., Hjelm, F., Bellgard, S. F., Bodley, E., & Jesson, L. K. (2022). Cross cultural leadership enables collaborative approaches to management of kauri dieback in Aotearoa New Zealand. *Forests, 12*, 1671. https://doi.org/10.3390/f12121671

Hunter, M. A., Aprill, A., Hill, A., & Emery, S. (2018). *Education, arts and sustainability: Emerging practice for a changing world.* London and Berlin: SpringerBriefs in Education.

Lambert, S. J., Waipara, N., Black, A., Mark-Shadbolt, M., & Wood, M. (2018). Indigenous biosecurity: Māori responses to kauri dieback and myrtle rust in Aotearoa New Zealand. In J. Urquhart, M. Marzano, & C. Potter (Eds.), *The human dimensions of forest and tree health: Global perspectives* (pp. 109–137). Cham: Springer International Publishing and London: Palgrave Macmillan.

Lee-Morgan, J. B. J. (2019). Pūrākau from the inside out: Regenerating stories for cultural sustainability. In J. A. Q. Q. Xiiem, J. B. J. Lee-Morgan, & J. D. Santolo (Eds.), *Decolonizing research: Indigenous storywork as methodology.* London and New York, NY: Bloomsbury Academic & Professional.

Lee, J. (2009). Decolonising Māori narratives: Pūrākau as a method. *MAI Review*, *2*, 1–12. https://www.review.mai.ac.nz/mrindex/MR/article/download/242/242-1618-1-PB.pdf

McEntee, M., Harvey, M., Mullen, M., Houghton, C., & Craig-Smith, A. (2023). Interweaving multiple knowledges to support children's participation and engagement in biosecurity and forest health: Toitū te Ngahere. *Knowledge Cultures, 11*(1), 154–183. https://doi.org/10.22381/kc11120239

McNaughton, M. J. (2004). Educational drama in the teaching of education for sustainability. *Environmental Education Research, 10*(2), 139–155. https://10.1080/1350462042000198140

McNaughton, M. J. (2006). Learning from participants' responses in educational drama in the teaching of Education for Sustainable Development. *Research in Drama Education: The Journal of Applied Theatre and Performance, 11*(1), 19–41. https://doi.org/10.1080/13569780500437572

McNaughton, M. J. (2010). Educational drama in education for sustainable development: Ecopedagogy in action. *Pedagogy, Culture & Society, 18*(3), 289–308. https://doi.org/10.1080/14681366.2010.505460

Ministry for Primary Industries. (2016). *Biosecurity 2025: Direction statement for New Zealand's biosecurity system.* Wellington: Ministry for Primary Industries Manatū Ahu Matua. Retrieved February 1, 2023, from https://www.mpi.govt.nz/dmsdocument/14857

Ministry for Primary Industries. (n.d.). https://www.mpi.govt.nz/forestry/new-zealand-forests-forest-industry/about-new-zealands-forests/

Nanson, A. (2021). *Storytelling and ecology: Empathy, enchantment and emergence in the use of oral narratives.* London and New York, NY: Bloomsbury Publishing.

Niemelä, R., Laulumaa, A. M., Tupala, A. K., & Raatikainen, K. J. (2023). A detour in research through the gorge: Approaching human–nature connections with site-specific performance. *Applied Theatre Research, 11*(1), 71–92.

Penetito, W. (2009). Place-based education: Catering for curriculum, culture and community. *New Zealand Annual Review of Education*, *18*, 5–29.

Ram, R, France, B., & Birdsall, S. (2016). Why biosecurity matters: students' knowledge of biosecurity and implications for future engagement with biosecurity initiatives. *Research in Science & Technological Education*, *34*(1), 69–84. https://doi.org/10.1080/02635143.2015.1066324

Raphael, J. (2021). Eco-anxiety and drama education. *Australian Drama Education Magazine: Drama and Sustainability*, *17*, 8–14. https://dramaaustralia.org.au/wp-content/uploads/2022/04/ADEM_No17_2021.pdf

Stats NZ. (2023). https://www.stats.govt.nz/news/our-indigenous-species-are-at-risk-of-extinction/

Toome-Heller, M., Ho, W. W. H., Ganley, R. J., Elliott, C. E. A., Quinn, B., Pearson, H. G., & Alexander, B. J. R. (2020). Chasing myrtle rust in New Zealand: Host range and distribution over the first year after invasion. *Australasian Plant Pathology*, *49*, 221–230. https://doi.org/10.1007/s13313-020-00694-9

Williams-Lemi, S., & Faau, A. (2021). *Guardians of Papatūānuku*. https://www.loopytunespreschoolmusic.com/wp-content/uploads/2022/02/Guardians-of-Papatūānuku

11 Socio-environmental (in)justice in forest stewardship

A multi-perspective materials assemblage for English as a foreign language teacher education in Chile

Katharina Glas and Miguel Farías

Introduction

As consumers, we are often faced with the choice between different products, one of which is praised as the most sustainable option. As citizens, we are challenged to be well-informed and take agency in various instances of democratic life. As living beings, we may sometimes feel overwhelmed with the pain we (un)knowingly inflict on other living beings, simply by taking part in the rituals and habitual practices of consumerist, throwaway societies. As teacher educators, we find that these roles intermingle in our attempts to project to future generations our desires for social justice and global ecological sustainability in the context of English as an international language in the Chilean curriculum. In this context, we understand language arts as encompassing critical literacy competencies for appropriately understanding and producing texts from diverse genres with awareness of multiple perspectives, audiences and purposes, unequal global power relations and the possibilities to transcend local contexts through digital media. In this sense, students expand the variety of semiotic resources available to them, enabling them to "participate in the world … looking at an issue … in different ways, analysing it, and suggesting possibilities for change and improvement" (Vásquez, Janks & Comber, 2019, p. 300).

The teaching proposal presented in this chapter has grown in Alimapu, the Mapuche name for the area around Valparaíso, Chile, referring to the hot, semi-arid climate of this land, hundreds of kilometres north of the forests that are the focus of this chapter. By making this toponymic reference, we acknowledge the role that Indigenous peoples have had in the stewardship of this land, past and present. In empathy, we share their anguish when travelling for hours through "green deserts" of introduced pine and eucalyptus plantations, alternating with clear-cuts, to reach the (still) pristine rainforests of the Lake District in Southern Chile. We also take clear sides in the ongoing conflict between (multi)national forestry companies and Mapuche Indigenous groups in the centre-south of the country.

Embedded in a national context, this proposal consists of a multi-perspective assemblage of authentic materials (Volkmann, 2010) and tasks to

DOI: 10.4324/9781032634524-15

foster critical ecoliteracy and problem-solving competencies designed for global English learners, in mother tongue or additional language contexts. With this genre- and content-based literacy project, we seek to integrate task- and project-based language learning with eco-literacy aims, proposing tasks that encourage meaningful reading and writing in English around important environmental issues. The selected materials have the potential to increase the visibility of global connections and contradictions that have led to socio-environmental injustices, by contrasting Indigenous perspectives on humans' role of stewardship of the natural world with those of a consumer-driven global economy that has normalised the existence of single-use disposable products. Thus, we hope to contribute to restoring caring relationships towards what Abram (1996) describes as the *more-than-human world*. Possible target audiences of this proposal are teenage and young adult learners in secondary or higher education settings, including English for Specific or Academic Purposes, especially in study areas such as forestry or environmental engineering. In this sense, this proposal is nourished by and can complement already existing proposals, including place-based ecopedagogy (for example: Eyraud, 2020), transepistemic approaches (Meighan, 2023), transcultural perspectives (Glas & Volkmann, 2021) and critical pedagogy approaches to English Language teaching (Crookes, 2021; Veliz & Veliz, 2019). For those who seek to go further, it can also provide the basis for project work facilitating interdisciplinary or international (multilingual) cooperation. In what follows, to frame the selection and assemblage of different authentic materials, we introduce the design of critical literacy tasks, how they relate to the concept of sustainable development and its competencies and the analysis of the responses gathered from a pilot study carried out as part of an environmental literacy project in teacher education at a state university in Santiago de Chile.

Critical literacy aims

As a proposal for English language arts learners around the globe, the presented tasks and materials aim to foster critical literacy through deep engagement with mainstream and Indigenous digital and/or multimodal texts, and a decolonising use of the English language. This links in with the development of research competencies, such as recognising different text genres, or locating specific information in long texts (see for example: Afflerbach & Cho, 2010; Rose, 2018). These competencies can be framed as tasks that integrate language learning and environmental "detective work" in the form of critical inquiry. Through strategic reading of the selected materials, students complete overarching discovery tasks together with specific learning outcomes:

- *Discovering shared responsibilities in malfunctioning forest stewardship.*
- Explain global interconnections in the production and consumption of single-use wood/paper products.

- Map global interconnections in forest stewardship.
- Explore mechanisms and consequences of structural and epistemic violence against Indigenous people and Indigenous knowledge on forest stewardship.

From there, they move on to a solutions-focused exploration task:

- *Exploring alternatives to single-use products made of wood and paper:*
  - Elaborate proposals for the replacement of single-use wood/paper products with more sustainable alternatives on various decision-making levels (for example: campus, local, corporate, regional, national).

After the initial work on receptive skills, students will be better prepared to continue with their expressive abilities to communicate their research results effectively. Some possible creative outcomes of the project could focus more on oracy work and include the creation of multimedia digital stories, audio-visual documentaries or multi-voiced, even multilingual drama projects (Beach, Share & Webb, 2017). On the other hand, persuasive writing skills could be practised in the composition of well-justified and researched petitions, practising *skilful argumentation* (Coiro, 2017; Fisher & Frey, 2013).

In the following section, we will outline how these literacy and oracy aims link in with UNESCO's Framework of Education for Sustainable Development Goals (2017). Then, we will describe the selection of the materials to be used for the four steps of this project. The full webquest can be accessed under caringlanguageteaching.org.

Education for sustainable development (ESD) – learning objectives and development of competencies

Our proposal is aligned with UNESCO's Framework of *Education for Sustainable Development Goals* (2017), due to its global scope, the variety of aspects considered for building a better future for the planet, as well as a clear breakdown of competencies to be fostered in educational settings. We are aware that this well-known and widely analysed document is based on an ambivalent, anthropocentric ecosophy[1] (Stibbe, 2021) and that it has received justified criticism (for example: Kopnina, 2020). The work with this proposal is therefore also an opportunity to enter a dialogue that allows us to evaluate the affordances and limitations of the UNESCO framework. The 2030 Agenda for Sustainable Development understands itself as "part of a global movement to eradicate poverty through 17 Sustainable Development Goals by 2030" (UNESCO, 2017, p. iv). These goals provide different themes to be considered in sustainability education, some of which are more closely related to environmental concerns, such as clean water and energy, sustainable cities and communities, responsible consumption and production, climate action, life below water and on land. Other goals envision sustainability regarding greater social justice: reduced inequalities (including gender), zero poverty and hunger, healthcare, education and working conditions, as well as structural preconditions for the attainment of the former, such as infrastructure, strong institutions and partnerships. The goals are complemented with eight "cross-cutting key competencies": Systems thinking, Anticipatory, Normative, Strategic, Collaboration, Critical thinking, Self-awareness and Integrated problem-solving competencies (UNESCO, 2017, p. 10). They aim at comprehending the connections between the 17 goals, as well as developing problem-solving capacities of *sustainability citizens* (Wals, 2015) that will (and are) inhabit(ing) an increasingly uncertain future.

For our proposal, linking in with the critical literacy aims described above, we identify three competencies as crucial, and expect students to generate evidence of their development (UNESCO, 2017, p. 10):

- *Systems thinking competency*: students will "recognize and understand relationships" between issues such as: global production and consumption patterns of disposable wood and paper products, un/sustainable forestry practices, the existence of international organizations accrediting sustainable forestry practices, poverty in the global south, (post)colonial conflict, as well as environmentally ambivalent and potentially misleading advertising strategies ("greenwashing") conveyed in discourse (Stöckl & Molnar, 2018) .
- *Critical thinking competency*: students will "question norms, practices and opinions" regarding sustainable and unsustainable forest stewardship; "reflect on their own values, perceptions and actions" with regard to consumption patterns of disposable wood and paper products; and "take a position in the sustainability discourse" by being able to relate and contrast it to indigenous views that propose a reciprocal attitude with nature.
- *Integrated problem-solving competency*: students will "apply different problem-solving frameworks to complex sustainability problems and develop viable, inclusive and equitable solution options", for example regarding the use of disposable wood and paper products.

(UNESCO, 2017, p. 10)

Further, the ESD Goals are formulated as specific learning objectives that cover three different domains: cognitive, socio-emotional and behavioural. For example, this proposal could be aligned with the following specific objectives regarding the "No Poverty" goal:

- Cognitive domain: "The learner knows about causes and impacts of poverty such as unequal distribution of resources and power, colonization … environmental degradation … and the lack of social protection systems and measures."
- Behavioural domain: "The learner is able to publicly demand and support the development and integration of policies that promote social and economic justice."

- Socio-emotional domain: "The learner is able to raise awareness about extremes of poverty and wealth and encourage dialogue about solutions."

(UNESCO, 2017, p. 12)

Similarly, our proposal shares some specific objectives with the ESD goal "Responsible Consumption and Production", such as being able to understand "the interrelatedness of production and consumption" (cognitive domain), "to encourage others to engage in sustainable practices in consumption and production" (socio-emotional domain) and "to challenge cultural and societal orientations in consumption and production" (behavioural domain) (UNESCO, 2017, p. 34). For a language arts project, it is important to note that socio-emotional and behavioural objectives often relate to effective persuasive or argumentative communication.

Forest Stewardship Proposal - materials selection and pedagogical uses

Step 1. Introduction to the topic

To begin, we initiate a brief thinking routine for students to analyse an image of a product advertisement about single-use wooden utensils (Image 11.1). The "See-Think-Wonder" routine encourages students to closely observe the image and notice details in the advertisement:

- What do you see?
- What do you think is going on?
- What does it make you wonder?

(Ritchhart, Church & Morrison, 2011, pp. 55–57)

In the ensuing "Thinking" part of the pedagogical routine, students are invited to phrase their interpretations such as pointing out possible contradictions, based on specific observations as supportive evidence – helping students avoid making unsubstantiated assertions. Finally, the "Wondering" step of the routine prompts students to ask broader questions, for example: "I wonder if these products are really more environmentally responsible than plastic products?"

For the following part, a juxtaposition of images and texts contrasts two opposing ideas: images depicting the treatment of the natural world as an exploitable, disposable resource, including a variety of disposable wood and paper products (for example: paper cups, wooden forks and spoons), as well as images of clear-cuts in monoculture forestry plantations, are opposed to texts in which students encounter Indigenous voices sharing their notions about the deep respect for other living beings ingrained in native cultures across the world. This resonates with an argument set forth by the New London Group: "When learners juxtapose different languages, discourse, styles, and approaches, they gain substantively in metacognitive and metalinguistic abilities and in their ability to reflect critically on complex systems and their interactions" (Cope & Kalantzis, 2000, p. 15).

We selected texts by two renowned Indigenous authors from different contexts. The first is Robin Wall Kimmerer, member of the Citizen Potawatomi Nation and State University of New York Distinguished Teaching Professor for Environmental Science and Forestry. In 2013, she published a widely celebrated collection of essays (*Braiding Sweetgrass: Indigenous Wisdom, Scientific Knowledge, and the Teachings of Plants*) offering deep insights into her ground-breaking work of introducing Indigenous knowledge on botany and environmental stewardship into the world of Western scientific research. In order to offer students a multimodal experience, we selected extracts from an interview with Kimmerer on the *OnBeing* podcast. It provides access to written interview transcripts, so students can read and listen to Kimmerer's voice at the same time. The second author is Elicura Chihuailaf, physician, poet and member of the Mapuche nation, the largest Indigenous group living on Chilean and Argentinian state territories. In 2020, Chihuailaf won Chile's National Prize for Literature for his poetry written in Spanish and Mapudungun (the Mapuche language). In order to make the inclusion of his texts meaningful for an English language arts or an English as a Foreign Language (EFL) class, we used extracts from interviews with him that were translated into English and published in the US-based journal *Latin American Literature Today.* For a multilingual language arts project, extracts from his essays in Spanish could be an enlightening addition.

The first extract from the *OnBeing* podcast (Tippett & Kimmerer, 2016) aims at students reflecting on the sustainability discourse and discovering joyful alternatives that assign humans a different role in their interaction with the Earth:

> [T]he language of sustainability is pretty limited. … [A]t its heart, sustainability … is embedded in this worldview that we, as human beings, have some ownership over these what we call resources, and that … human beings can keep taking and keep consuming. The notion of reciprocity is really different from that. It's an expansion from that, because what it says is that our role as human people is not just to take from the Earth, and the role of the Earth is not just to provide for our single species … not only does the Earth sustain us, but … we have the capacity and the responsibility to sustain her in return. So it broadens the notion of what it is to be a human person, not just a consumer. And there's such joy in being able to do that, to have it be a mutual flourishing instead of the more narrow definition of sustainability so that we can just keep on taking.

In a second extract from the same source, Kimmerer explains how the English language is limited to using "it" for beings like sugar maples or salamanders. In her perspective, this feels as disrespectful as calling your own grandmother "it". In Potawatomi, instead, animate and inanimate cases exist: "It is impossible in our language to speak of other living beings as 'its'" (Tippett & Kimmerer, 2016). This can create critical intercultural language awareness by calling attention to and comparing grammatical features of different languages, and how they embody respect or disrespect for the natural world. Optionally, this extract could lead on to a brief discussion about the Sapir-Whorf hypothesis: up to which point do the cognitive perceptions created by our native languages permit us to overcome barriers in our thinking and embrace new ways of perceiving and interacting with the environment?

The third extract, taken from the interview with Chihuailaf, allows learners to move on from the previous reflection by inviting all readers (not only Indigenous ones) to "wake up" to the memory held alive by native peoples regarding the oneness of all life on earth. At the same time, it creates a parallel structure between fragmentation and violence within a global system of inequalities:

> [A]ll of us who live in this world come from native peoples / cultures, but only a small number of us still maintain the memory of our belonging to the Earth, to Nature, no more and no less than all living beings and all those that appear inanimate, like stones and minerals. It is the awakening of deep nations and peoples – all over the planet – that are beginning to realize we have to fight against fragmentation, against the violence carried out by a small group of families that have taken hold of power and embedded themselves within it (the superficial United States; the superficial, alienated Chile). *Itrofil mogen*: the whole without exclusion, the entirety with no fragmentation of life, of all living things
>
> (Dixon & Chihuailaf, 2020)

Students read and discuss the three texts in small groups and are then prompted to write brief reflective responses on the impact that the juxtaposition of images and text has made on them.

Step 2. Exploring principles and criteria for forest stewardship

For the following step, students are invited to interact with a document entitled *FSC Principles and Criteria for Forest Stewardship* (2023). After the initial observation-thinking routine involving the image of the wooden utensils box (Figure 11.1), students' attention is drawn more closely to the logo of the FSC certification and the name of the Forest Stewardship Council. A first question – "What does *Forest Stewardship* mean?" – leads to finding dictionary definitions of the word "stewardship" with its double connotation of "care" and "management" and discussing how both ideas can apply to humans' relationship with forests. Here, juxtaposed images of intact ancient forests and monoculture tree plantations further help to lead the discussion.

Figure 11.1 A thought-provoking lesson starter

In order to lead students towards autonomous exploration and inquiry, a useful new thinking routine could be applied here: Think-Puzzle-Explore (Ritchhart et al., 2011, p. 71), a modified version of the Know-Want to Know-Learned (KWL) routine. The following questions help students develop curiosity about the topic and formulate their own questions, which can then guide their interaction with the document:

- What do you think you know about the Forest Stewardship Council and its certifications?
- After the first introduction to the topic, what puzzles you about it? What questions do you have?

The questions are shared among the whole class. Students are then given access to the document and encouraged to look at the contents page before beginning to browse it. In this moment, students are asked to respond to the third question of the routine:

- "How might you explore the puzzles we have around this topic?"

A model puzzle question shared by the teacher might be: "What is the role of the *Forest Stewardship Council* and its international certifications in the protection of ancient forests and the communities that have traditionally cared for them?"

Finally, students can explore the document further to see which of the described principles may respond to the questions they asked. If questions remain, they may be included in the following step.

Step 3. "Detective work" on a socio-environmental conflict: a critical inquiry (webquest)

Step 3 is designed as an open-ended group inquiry task, in which students interact with a variety of different sources (Fisher & Frey, 2013) that shed light on the conflict between Mapuche communities in Southern Chile and multinational forestry companies. Sources are accessed through a webquest linking documents and websites, combined with questions that students are to explore through strategic, selective reading. For language work, an interesting fact is that because of the international certifications for forestry companies in Chile, the documents are published online in the *lingua franca* English.

1) "Letter to FSC for complicity in crimes and impacts of industrial forestry exploitation in Chile": An open letter on the *World Rainforest Movement* website, (World Rainforest Movement, 2017) signed by thirty organisations.
2) A report commissioned by the FSC and written by Rosamel Millaman and Charles Hale (2016): *Chile's Forestry Industry, FSC Certification and Mapuche Communities.*
3) An academic article by a team of Chilean and Mapuche social scientists published in the journal *Human Ecology*, with the question "Do Large-Scale Forestry Companies Generate Prosperity in Indigenous Communities?" as its title (Hofflinger, Nahuelpan, Boso & Millalen, 2021).
4) The CMPC website[2] (CMPC, n.d.) and its section on the "environmental dimension": a global forestry company online.

Although students are encouraged to formulate their own questions as they go along, they can distribute the following six questions in their groups and gradually resolve them through collaborative inquiry.

1. What are the conflicts underlying the publication of the open letter that accuses the FSC of complicity in crimes and impacts of industrial forestry exploitation?
2. In what ways are the principles of the FSC in possible conflict with the forestry model in Chile?
3. What aspects of Mapuche *rakizuam* (thinking) in relation to the environment are in conflict with the forestry model in Chile?
4. How are texts in English as a *lingua franca* used as a means to alert international public attention to these conflicts?
5. Does large-scale forestry in the form of tree plantations contribute to socioeconomic justice? Do the different texts share the same opinion on this matter? Which of the texts seems to give the most reliable information in this respect?
6. Debate: Can large-scale forestry and tree plantations contribute to environmental sustainability? Support your opinions with evidence from the different texts.

The questions prompt students to collect evidence to gradually build up their arguments so they can take a position based on research, while learning to locate information in multiple sources,

Step 4. Taking position, taking action

The inquiry work done in the previous steps should culminate in a concrete proposal that can help a wider audience make conscious decisions regarding the use of single-use wood and paper products, taking into account that individual actions have a much smaller impact than decisions taken on more collective levels (for example: a school, a university campus, a municipality, a company). Depending on class sizes and the setting in which the proposal is implemented, students can elaborate proposals for sustainable alternatives to single-use wood and paper products, especially reusable utensils, such as washable cups and cutlery. The genres of these proposals may include informative posters, well-supported petitions, discussion panels, short documentaries, reels – and more creative or artistic projects, such as dramatic representations, poetry or song writing, or collages combining text and image. One of the central principles of project-based work is to involve students in the decision-making process about the outcome and format of a "final product". The ideas above should, therefore, be shared with students only in order to give them inspiration and some initial guidance. In each learning context, there will be differences in the perceived relevance and feasibility for students to take action. Students' work in this part also requires teacher guidance. Apart from providing feedback on the quality of arguments and language, teachers can share model texts, such as the Canadian document *Guidance for Selecting Alternatives to the Single-Use Plastics in the Proposed Single-Use Plastics Prohibition Regulation* (Environment and Climate Change Canada, 2021). Here, students can find useful expressions that can be applied to argue for alternatives to single-use wood and paper products.

Responses from the pilot study

The materials were used in an elective environmental literacy class in an initial teacher education programme for English as an Additional Language (EAL) in Chile. After an introductory lesson, students were invited to provide informed consent to participate in the study. Autonomous access to the materials was given, together with an invitation to respond to the questions in the proposal. The responses were recorded in students' ecojournals, where they wrote regular narrative entries on assigned topics discussed in the elective course as part of assessed coursework. Students' responses were analysed according to the ESD and critical literacy goals outlined earlier.

Reacting to the juxtaposition of the texts by Robin Wall Kimmerer and Elicura Chihuailaf, with the images of disposable wooden forks, an intact forest and a clearcut, one student draws attention to three concepts needed in today's world that may not yet have been sufficiently included in the sustainability debate: "Kimmerer and Chihuailaf emphasised the importance of reciprocity, interconnectedness, and respect for the Earth. I consider these are all important principles that we should keep in mind as we seek to adapt to a more sustainable way of living". Her response shows clear evidence of critical thinking, referring to the presented contradictions: "It is ironic how something that is supposed to be good (wooden utensils), still causes harm (cleared forests)".

With respect to the power of language to describe what is considered animate, and its repercussions on the relationships of exploitation we establish with nature, another student writes:

> In a way, English kidnaps the terms of equality for non-human beings, such as animals, insects, and nature, to be replaced and treated as mere things that can also be possessed. Then, as a tree is a thing that can be possessed, little thought and remorse take place [sic] before transforming that tree into a single-use and disposable piece of cutlery.

The following quote shows that reading about forest stewardship can help students develop cognitive, socio-emotional and behavioural learning objectives regarding a deeper understanding of forest ecology and its contradictions, with forestry practices posing a threat to healthy soils. This new knowledge is the basis for the ability to argue and raise one's voice for sustainable practices, as well as getting involved in collective action and international campaigning (see UNESCO, 2017, p. 40, SDG 15 "Life on Land"):

> Before knowing this information, I would've thought that forestry and tree plantations would always contribute to environmental sustainability as I wasn't too well informed about different procedures that can be done in forestry. It makes me realise, once again, the importance of being informed and aware of different environmental issues and practices in order to make a contribution instead of damaging when trying to help and preserve the environment.

At the same time, critical language awareness is developed when students can see English as a lingua franca "in action" used for international campaigns shown in some of the selected texts, as expressed in the following quote:

> In the case of the FSC conflict with the Mapuche community, the use of English as a Lingua Franca serves as a medium to make visible to the international community that this worldwide corporation that claims to stand for the stewardship of the forests is certificating [sic] forestry companies that are closely linked to human right violations against indigenous communities, tax evasion and misuse of natural resources, which is a huge contradiction.

Feelings of unease and hope, as well as reflections about the distribution of power and the possibility to organise collectively to challenge unequal structures, are clearly located in the socio-emotional domain of the SDG learning objectives. They are shared by another student:

> It surprised me that organisations such as the FSC have a huge impact on our forests and land, and how important it is that these organisations are kept in check and make sure to take care of nature and natural resources. It makes me feel somewhat uncomfortable that these big corporations have so much power and as citizens we have not got many things to do to help. The fact that a community such as the Mapuche community can come together, on the other hand, makes me feel hopeful for the changes that are actually possible to make.

Concerning the question about what aspects of Mapuche *rakizuam* (thinking) in relation to the environment are in conflict with the forestry model in Chile, students are able to explain this conflict in their own words:

> There are several aspects of Mapuche *rakizuam* in conflict with the forestry model in Chile. On the one hand, the Mapuche people view the land as sacred and believe in a profound spiritual connection with their ancestral territories. For them, the land is not merely a resource to be exploited, but a living entity with intrinsic value and cultural significance. And the forestry model often disregards this spiritual and cultural connection, leading to conflicts over land use and ownership, because we know the forestry model promotes large-scale tree plantations and logging.

After understanding the causes of the conflicts more deeply, students can also make proposals about how to manage environmental conflict better:

> In my opinion, it is terrible how a company that must ensure care for the environment, damages it and violates indigenous people's rights, which

makes us doubt about the companies in charge of caring for nature. In this sense, there should be an involvement of the Mapuche community, since they are one of the communities that is most concerned about the environment. Hence, there should be equitable approaches in the management of natural resources, recognizing the cultural significance of indigenous communities and their connection with land.

Conclusion

Analysing our students' responses, we can see that the materials we assembled, together with the tasks aligned with the ESD goals, allow learners to develop a deeper understanding and a critical stance towards socio-environmental issues in a globalised world. The ESD goals provide an ample frame of reference for critical environmental education grounded in local communities. They help us formulate clear cognitive, socio-emotional and behavioural environmental learning objectives while teaching English as a lingua franca. However, the recent escalation of the climate emergency towards a phase of so-called *global boiling* requires holistically educated citizens to engage in urgent action. This forces us to sharpen our vision to be able to identify important limitations in the ESD framework. Anthropocentric discourses, including the ESD goals, continue to propagate the view that as humans, we primarily *take from* the earth, treating her gifts as resources to be owned and used, however carefully. Conversely, ecocentric discourses give us the opportunity to see our role as *both receiving and giving*. Kimmerer develops the principle of *reciprocity*, rooted in traditional ecological knowledge, to renew environmental thinking and education from an indigenous perspective. A reciprocal relationship with the earth includes spiritual (gratitude, mindful consumption) as well as practical aspects (civic engagement, land stewardship, and ecological restoration). It also entails the gift of the use of language and stories:

> We may not have wings or leaves, but we humans do have words. Language is our gift and our responsibility. I've come to think of writing as an act of reciprocity with the living land. Words to remember old stories, words to tell new ones, stories that bring science and spirit back together to nurture our becoming people made of corn.
>
> Kimmerer (2013, p. 347)

Notes

1 The term ecosophy, coined by Naess (1986) to describe a philosophy of deep ecological harmony, is used by Stibbe (2021) in a broader sense, to circumscribe different schools of thought around ecological issues, ranging from anthropocentric, neoliberal and/or optimistic approaches to ecocentric, socialist and/or pessimistic ones. In this sense, discourses can be destructive, ambivalent or beneficial depending on the ecosophy adopted by a researcher.

2 CMPC stands for *Compañía Manufacturera de Papeles y Cartones*, the privately owned Chilean manufacturing company of papers and cardboards.

References

Abram, D. (1996). *The spell of the sensuous: Perception and language in a more-than-human world*. New York, NY: Pantheon.

Afflerbach, P., & Cho, B.-Y. (2010). Determining and describing reading strategies: Internet and traditional forms of reading. In H. S. Waters & W Schneider (Eds.), *Metacognition, strategy use, and instruction* (pp. 201–225). New York, NY: Guilford Press.

Beach, R., Share, J., & Webb, A. (2017). *Teaching climate change to adolescents: Reading, writing, and making a difference*. New York, NY: Taylor & Francis.

CMPC. (n.d.). Environmental dimension. https://www.cmpc.com/en/sustainability/#ambito-ambiental

Coiro, J. (2017). Advancing reading engagement and achievement through personal digital inquiry, critical literacy, and skilful argumentation. In C. Ng & B. Bartlett (Eds.), *Improving reading and reading engagement in the 21st century: International research and innovation* (pp. 49–76). Singapore: Springer.

Cope, B., & Kalantzis, M. (Eds.). (2000). *Multiliteracies: Literacy learning and the design of social futures*. New York, NY: Psychology Press.

Crookes, G. V. (2021). Critical language pedagogy: An introduction to principles and values. *ELT Journal, 75*(3), 247–255. https://doi.org/10.1093/elt/ccab020

Dixon, M., & Chihuailaf, E. (2020, November). "Within each one of us lives the thought of our elders and our ancestors": A conversation with Elicura Chihuailaf. *Latin American Literature Today*. https://latinamericanliteraturetoday.org/2020/11/within-each-one-us-lives-thought-our-elders-and-our-ancestors-conversation-elicura/

Environment and Climate Change Canada. (2021). *Guidance for selecting alternatives to the single-use plastics in the proposed single-use plastics prohibition regulation*. https://www.canada.ca/en/environment-climate-change/services/managing-reducing-waste/consultations/proposed-single-use-plastics-prohibition-regulations-consultation-document.html

Eyraud, K. E. (2020). A place-based ecopedagogy for an English for academic purposes program. In J. Goulah & J. Katunich (Eds.), *TESOL and sustainability: English language teaching in the anthropocene era* (pp. 111–131). London: Bloomsbury.

Fisher, D., & Frey, N. (2013). Reading and reasoning: Fostering comprehension across multiple texts. http://www3.canisius.edu/~justice/CSTmodule-final/ira-essentials-8026-reading-and-reasoning.pdf

Forest Stewardship Council (FSC). (2023). *FSC principles and criteria for forest stewardship*. https://connect.fsc.org/document-centre/documents/resource/392

Glas, K., & Volkmann, L. (2021). Ecocriticism meets postcolonialism: Interrelating two critical perspectives for the EFL classroom. In C. Ludwig & C. Deetjen (Eds.), *The world beyond: Developing critical environmental literacies in EFL* (pp. 41–64). Heidelberg: Winter.

Hofflinger, A., Nahuelpan, H., Boso, À., & Millalen, P. (2021). Do large-scale forestry companies generate prosperity in indigenous communities? The socioeconomic impacts of tree plantations in Southern Chile. *Human Ecology, 49*(5), 619–630. https://doi.org/10.1007/s10745-020-00204-x

Kimmerer, R. W. (2013). *Braiding sweetgrass: Indigenous wisdom, scientific knowledge and the teachings of plants*. Minneapolis, MN: Milkweed.

Kopnina, H. (2020). Education for the future? Critical evaluation of education for sustainable development goals. *The Journal of Environmental Education*, *51*(4), 280–291. https://doi.org/10.1080/00958964.2019.1710444

Meighan, P. J. (2023). Transepistemic English language teaching for sustainable futures. *ELT Journal*. http://dx.doi.org/10.1093/elt/ccad004

Millaman, R., & Hale, C. (2016). Chile's forestry industry, FSC Certification and Mapuche Communities. https://ga2017.fsc.org/wp-content/uploads/2017/10/2017_04_02_-Final-Report.pdf

Naess, A. (1986). Self-realization: An ecological approach to being in the world. Reprinted in G. Sessions (Ed.), (1995). *Deep ecology for the twenty-first century* (pp. 225–239). Boston: Shambhala Publications.

Ritchhart, R., Church, M., & Morrison, K. (2011). *Making thinking visible*. San Francisco, CA: Jossey-Bass.

Rose, D. (2018). Languages of schooling: Embedding literacy learning with genre-based pedagogy. *European Journal of Applied Linguistics*, *6*(1), 59–89. http://dx.doi.org/10.1515/eujal-2017-0008

Stibbe, A. (2021). *Ecolinguistics. Language, ecology and the stories we live by* (2nd ed.). Abingdon and New York, NY: Routledge.

Stöckl, H., & Molnar, S. (2018). Eco-advertising: The linguistics and semiotics of green (-washed) persuasion. In A. F. Fill & H. Penz (Eds.), *The Routledge handbook of ecolinguistics* (pp. 261–276). Abingdon and New York, NY: Routledge.

Tippett, C., & Kimmerer, R. W. (2016). The intelligence of plants. *On Being* [Podcast]. https://onbeing.org/programs/robin-wall-kimmerer-the-intelligence-of-plants-2022/

UNESCO. (2017). *Education for sustainable development goals. Learning objectives*. https://unesdoc.unesco.org/ark:/48223/pf0000247444/PDF/247444eng.pdf.multi

UNESCO. (2021). *UNESCO and sustainable development goals*. https://en.unesco.org/sustainabledevelopmentgoals

Vásquez, V. M., Janks, H., & Comber, B. (2019). Critical literacy as a way of being and doing. *Language Arts*, *96*(5), 300–311. https://doi.org/10.58680/la201930093

Veliz, L., & Veliz-Campos, M. (2019). An interrogation of the role of critical thinking in English language pedagogy in Chile. *Teaching in Higher Education*, *24*(1), 47–62. http://dx.doi.org/10.1080/13562517.2018.1456424

Volkmann, L. (2010). *Fachdidaktik Englisch: Kultur und Sprache*. Tübingen: Narr.

Wals, A. E. J. (2015). *Beyond unreasonable doubt. Education and learning for socio-ecological sustainability in the Anthropocene*. Wageningen, Wageningen University. Retrieved July 26, 2023, from https://arjenwals.files.wordpress.com/2016/02/8412100972_rvb_inauguratie-wals_oratieboekje_v02.pdf

World Rainforest Movement. (2017, November 6). Letter to FSC for complicity in crimes and impacts of industrial forestry exploitation in Chile. https://www.wrm.org.uy/action-alerts/letter-to-fsc-for-complicity-in-crimes-and-impacts-of industrial-forestry-exploitation-in-chile

Part 4

Engaging with visual and social media

12 The visual image and imagination of alternative futures

Nuance, ambiguity and hope

Peter Sutoris

Introduction

The importance of visual images in our culture is difficult to overstate. From the billions of images shared on social media and millions of hours of video content streamed every day to the ubiquity of visual image-driven marketing, images are at the heart of our civilisation, and spaces of education are no exception. Countless images appear in textbooks, presentation slides, videos and animations used across physical and virtual learning environments. Some of these images are imbued with layers of intentional meaning, while others are "mere" illustrations of the words they accompany. Whether intentionally or not, images inform, unsettle, challenge, reaffirm and redefine our understandings of the world we inhabit – our environment and our place in it – and therefore also how we think and feel about the future. Whether we like it or not, we live in a soup of visuals.

Despite the ubiquity of images and their crucial importance in shaping our identities, understandings and imaginations, many of the world's education systems remain firmly committed to focusing on the written word as a locus of meaning and learning. Much time and effort goes into reading, writing and understanding words, and it is verbal expression that forms the backbone of assessment systems propping up the credentialism of contemporary education (cf. Brown, 2001). Save for specialised spaces of learning in fields like art history or film studies, far less time and effort is typically dedicated to observing, creating, reading, analysing and reflecting on visual images. If educators are unconcerned with whether we sink or swim in the visual soup around us, they help render visual images "neutral" in their effects on individuals and society at large.

But visual images are anything but neutral. This is the point underpinning the idea of a "visual diet" (Tanguy, 2021). Defined as "the visual content that a person willingly or unwillingly consumes per day" (*Visual Diet*, n.d.), the concept is at the heart of a campaign that has called for a rethinking of society's collective relationship with the visual in its manifesto:

DOI: 10.4324/9781032634524-17

> We are what we see and, at the moment, we're being visually poisoned. Bullied by perfect selfies and ignorant stereotypes. Oppressed by images that make us feel inadequate and depressed. We need to open our eyes to the importance of what we see. And curate a more varied, more nourishing diet. Feast our eyes on things we've never experienced before.
>
> (*Visual Diet*, n.d.)

While this campaign is situated primarily within the context of debates around the proliferation of harmful social media images and associated harms to mental health, its message has a wider relevance. The underlying idea that visual images have consequences to which we need to be sensitised matters to conversations about education's role in contributing to and addressing environmental decay.

Within the vast field of scholarship about Environmental and Sustainability Education (ESE), I have positioned my work both as an educator and as a researcher primarily around the idea of imagination of alternative futures. It is with respect to this idea that this chapter discusses the significance of the visual image, motivated by questions like: How might visual images stimulate the imagination of alternative futures? In what ways can visual images get in the way of radical imagination? What is the role of visual literacy in ESE? The chapter begins with a brief discussion of the link between education and imagination of alternative futures before proceeding to discuss the ways that observing and consuming the countless images we encounter – the "visual soup" that surrounds us – might interact with our imaginations. This is followed by a discussion of how participating in "cooking" the "visual soup" might further help young people take control over how images affect their imaginations and concluding reflections about the importance of curating a visual diet. In offering these thoughts, I reflect on examples of my own work as a filmmaker, lecturer and researcher in which I have relied on visual images. I seek not to highlight my work or to claim it in any way superior to the work of others, but simply to use it as an opening for thinking through the many ways visual images impact the imagination of alternative futures and therefore the need to be mindful and intentional in their use. The discussion is motivated by the underlying argument that visual images matter and deserve greater focus and scrutiny by education practitioners and researchers alike, particularly in the context of unprecedented environmental decay that much mainstream education is, so far, largely failing to confront.

Education and Imagination of Alternative Futures

In popular and academic discourse alike, education is generally viewed as a force for good. In the context of debates about the environmental multi-crisis (Litfin, 2016), education has emerged as one of the solutions, an instrument in humanity's toolkit for tackling climate change and other forms of environmental decay (see for example: Anderson, 2012). Ideas like "climate change education" and "climate literacy" abound in the field of ESE, highlighting education's potential to help mitigate, if not completely transform, the multi-crisis (Reid, 2019).

What is typically absent from these accounts of education-fuelled brighter tomorrows is a recognition of education's historical *contribution* to the mess humanity finds itself in today. The ways in which formal education systems have played a role in breaking the bonds between people and the land they live on (such as in the case of residential boarding schools that removed Indigenous children from their communities), spreading notions of "success" and "the good life" rooted in logics of consumerism and capitalist accumulation, and to putting blinkers on young people's criticality (sometimes in the name of "critical thinking"!), are generally absent in this discourse. Scholars who have explored education from the perspective of Indigeneity (Prakash & Esteva, 2009), ecopedagogy (Misiaszek, 2016) and deschooling (Illich, 1971), among others, have pointed to these links, but we would be hard-pressed to find their insights reflected in policy frameworks advocating education as a crucial element of planetary salvation (see for example: Ehlers, Husain, Karboul, Kwauk, Lovett, McBride, McCabe, Steer & Vaduganathan, 2022).

These critical perspectives bring into question the extent to which educational efforts underpinned by mainstream discourses can find themselves "on the right side of the barricade" in this historical moment. As Smith (2005) has pointed out, contemporary schools often aim to produce "good environmental citizens" who are, in the context of neoliberal discourses around sustainability, broadly speaking, obligated to comply in a largely "apolitical" manner with "behavioural norms that facilitate the continuance of the current social/political system" (p. 51). Holfelder (2019) has similarly argued that much mainstream education conceptualises the idea of the future as "closed" or non-malleable, leading young people down the path of adapting to the changing circumstances rather than seeking to fundamentally transform the future.

My own research, both in schools and in non-formal spaces of learning among communities of environmental activists, has echoed these arguments, and led me to propose the idea of "educating for the Anthropocene" as an alternative conceptualisation of education in the context of the multi-crisis. This framework recognises education's historical responsibility for contributing to environmental decay in the context of the multi-crisis[1] and seeks to chart a path forward that learns from this history. One of the ingredients of "educating for the Anthropocene" that I have proposed in my earlier work is the idea of radical imagination of alternative futures:

> Being able to reimagine different worlds is the opposite of bureaucratization – it requires creativity and autonomy and relies on an internal locus of possibility. It defies predictability and planning, allowing access not just to the "here and now" but also the "there and then," the "not yet there," the "elsewhere," or the "nowhere" (Jovchelovitch,

> Priego-Hernandez & Glăveanu, 2017, p. 116). The high Anthropocene moment calls for a reimagining of the world that may be unthinkable to many. Just as "critical thinking" as a "twenty-first-century skill" has been co-opted in the service of neoliberal, technocratic future-making, so has imagination become constrained to a limited set of possibilities.
> (Sutoris, 2022, p. 236)

One of the ways that our society and culture constrain our imagination is the immersion in a visual soup of allegedly neutral images. Just as images have the power to open up new worlds, stimulating thinking and feeling about alternatives, they also have the ability to close down possibilities and preserve the status quo, as the examples in the next section demonstrate.

Swimming in the visual soup

Images affect our imagination of alternative futures in many different ways. They might convey new information, which may or may not be true. They may raise and answer questions. They may elicit an emotional response, such as feelings of care, anxiety about personal and collective futures, or disgust and anger at the ease with which the natural environment is being destroyed. They might fuel and reaffirm the viewer's resolve to act or undermine the viewer's motivation for pro-environmental action. They might support healthy, adaptive ways of responding to the environmental multi-crisis or they might contribute to the mental health epidemic associated with ecoanxiety and ecoparalysis (Palinkas & Wong, 2020). These different responses and outcomes contribute to "closing down" or "opening up" the future, stimulating or sedating the radical imagination. Different visual images will impact different people differently depending on the cultural, social and historical context, personal identity, prior experience, the presentation of the image with respect to other textual, visual or acoustic elements and a myriad of other factors. While the visual soup arguably exerts a large influence on the radical imagination, the impact of any individual image – as is the case with any cultural artefact – is often impossible to ascertain.

One of the ways in which scholars have tried to make sense of this complexity is through the concept of visual literacy. Many different definitions have been proposed since the idea was first coined by John Debes in 1969; what many of them have in common, as Kedra (2018) has shown in a review of visual literacy scholarship, is visual reading and writing skills. The reading skills include the ability to understand symbolic meanings, identify the intention and evaluate the meanings behind a visual image (Kedra, 2018). Visual literacy theory mirrors many of the concepts found in verbal literacy, including the idea of visual grammar, syntax and semantics (Felten, 2008). While debates about visual literacy have been going on for over half a century (and indeed

Figure 12.1 The contrasts of climate change discourse (courtesy of Pixabay)

much before that, as Peña and Dobson (2021) have pointed out), the idea has received little attention within most education systems, in line with the myth of the "neutral" visual image.

Applying the idea of reading an image to visuals circulating in conversations about environmental decay is a first step in understanding the ways in which these images might be impacting the radical imaginations of learners. Many of these images are intended to awaken the viewer to the stark realities of climate change through contrasting visions of the before and after, as seen in Figures 12.1 and 12.2. In both these images, the viewer confronts a dystopian vision of the future, represented by a barren, dried-out landscape (Figure 12.1) and a literally burning planet Earth (Figure 12.2). Both images rely on a symmetrical composition split between the "natural" and "unnatural" states of the environment. This juxtaposition could be interpreted as symbolising an underlying teleology of climate change, a linear progression over time from harmony to disharmony, from life to death. Particularly in Figure 12.2, the contrast is heightened by a hyperbolic depiction of the future, where the Earth dissolves into space, an outcome beyond even the most pessimistic of scientific predictions. Both images appear to be designed to elicit an emotional response – fear, horror, anger – as they highlight the urgency and magnitude of the issue at hand.

Given the long and troublesome history of climate denialism, it is not surprising that many visual depictions of climate change seek to shock the viewer

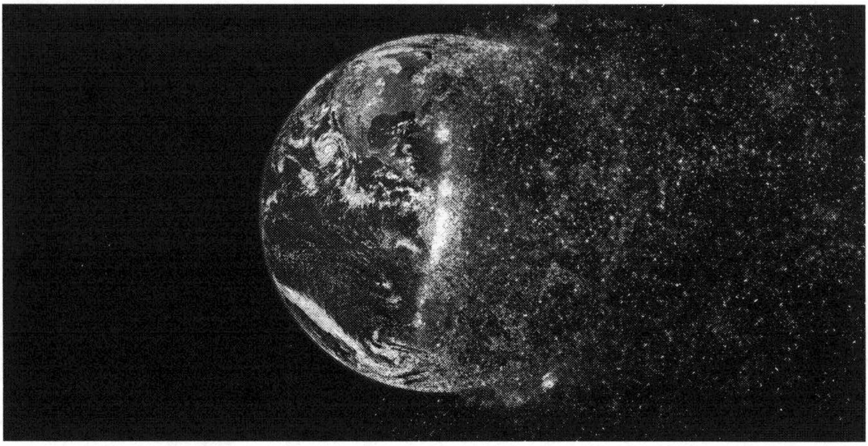

Figure 12.2 The apocalyptic visions of climate change discourse (courtesy of Pixabay)

out of complacency. If humanity is sleepwalking into disaster, it makes sense that those concerned about the future might pepper the visual soup with images intended to wake us up. Yet, it is not clear that this is the effect such images would have on the viewer. The exaggeration might be off-putting to some observers who might conclude that these images represent propaganda rather than a reflection of a scientific consensus. The absence of human elements might help reinforce the people-environment dualism, as if it was "the planet" that was at stake rather than the survival of human civilisation. (We have every reason to believe that "the planet" will indeed survive the current environmental multi-crisis, having survived many other disruptive events in its long history, but the survival of humanity is less clear.) Given that the idea of intrinsic value of nature may not be as widespread among those exposed to the image as a concern for personal well-being and the well-being of those close to us, the images might fail to elicit perhaps the most important of affective responses when it comes to climate change – care. Some might also view them as presenting a fatalistic vision of the future, with no scope for change and human agency. Such images could, therefore, end up contributing to anxiety and paralysis rather than to reflection and action.

Other images might be underpinned by a subtler approach. Figure 12.3 is an illustration depicting the impact of plastic pollution on life in the oceans. By showing fish freely swimming in the water alongside their counterparts trapped by plastic packaging, it too establishes a contrast between the "natural" and the "unnatural". But this is a contrast within a single moment of time rather than a before-after binary. Rather than conveying a sense of impending doom, it illuminates an unfolding process of change with no certain outcome. By zeroing in on a singular ecosystem and a singular source of pollution, it nudges the viewer towards a particular course of action, such as researching

Figure 12.3 Picturing the impact of plastic pollution (courtesy of Pixabay)

the impact of plastics on aquatic life or joining a campaign to ban plastic packaging. The depiction of specific animal species might elicit feelings of empathy in the viewer, helping to fuel action.

The point is not that some images are more effective or more appropriate than others. It is, rather, that the visual language deployed by each image interacts differently with pre-existing beliefs, recognisable symbols, shared affective states and cultural expectations within different settings. Images – particularly ones with "transparent" messaging and utilising a didactic visual language – are also underpinned by different intentions of their authors, and being attuned to the language they deploy can illuminate these intended effects, as well as help identify instances where images might miss their target. Being able to read an image is an important skill for anyone trying to swim in the visual soup, and it is even more important for educators wishing to be intentional about their practice. This is where I turn in the next set of reflections.

Ambiguity, atmosphere and grief

In my work as a filmmaker, ethnographer and lecturer, I often deal with visual images. They help me understand the realities unfolding in front of me, process my own emotions and share these experiences with my students. I had the most powerful experience of confronting, and teaching about, environmental degradation through the visual image in the Marshall Islands, while working on my documentary *The Undiscovered Country* (2012). A low-lying country of atolls in the Pacific, the Marshall Islands are among the world's territories most threatened by rising sea levels. They are also a former colonial territory of Japan, Germany and the United States, a site of Cold War era nuclear bomb testing by the United States, which irradiated thousands and vaporised entire islands (Barker, 2013), and one of the world's most foreign aid-dependent countries. With double epidemics of diabetes and malnourishment, most young people facing unemployment, and the beaches in this "tropical paradise" disappearing under the crashing waves, the fate of the Marshall Islands raises visceral questions about the ultimate end-point of the world's pursuit of "progress" and "development". It is a place where it is very easy to imagine the future as a closed one, to give into despair at the prospect of the entire world following down the path towards what David Wallace-Wells has called *The Uninhabitable Earth* (2019).

What I hoped to convey with the images I captured here was, however, something else. My goal was to help the viewer experience what I was experiencing, to enable them to be transported into a different reality through the images. I did not strive to convey any particular message. This does not mean that the images were somehow objective or free of my viewpoint. My response to what I was witnessing, which coloured the kinds of images I chose to capture, was a sense of loss and grief, a kind of "nostalgia for the future" (to quote the title of the film by Kishore & Karmakar, 2017). As Laura Ogden reminds us, "[l]oss is a refrain of the present. Much of the discussions about climate change and rates of species extinctions presents these phenomena as global and therefore universally distributed in space and time" (2021, p. 6). But this is not so. "Instead, loss is expressed in dialects and includes profound silences", Ogden writes. "Loss has its own vernaculars and place-based temporalities, even though these differences are often effaced by modernity's monocular optics" (ibid). Capturing the place-based loss in the Marshall Islands was what I was after. To do this, I looked for visual metaphors evocative of the sense of grief I sensed around, and within, me. It was not cerebral, fact-driven images that I hoped to produce but ones conveying a sense of atmosphere, ambiguous images to which different observers might respond differently – images that would help stimulate the radical imagination rather than suppress it.

The beaches of the Marshall Islands contained a surprising number of car wrecks, which caught my eye. One such wreck made for a particularly evocative contrast with a ship sailing away on the horizon (Figure 12.4), while another, abandoned inside a thick forest, lent itself to the question "a journey

Figure 12.4 Looking for a visual metaphor for progress (courtesy of Peter Sutoris)

to where?" (Figure 12.5). I also noticed crumbling tombstones along the beaches (Figures 12.6–12.7), and learned from local people that these were not washed ashore but were in fact the remnants of a cemetery that had been flooded due to rising sea levels. The local residents explained to me that their ancestors would have never buried their dead in a place where water could reach them, which demonstrated not only the extent of the sea level rise but also the many different ways in which it threatened the atoll's inhabitants and their heritage. To me, these images raised questions of freedom and captivity, of nature and industrial modernity, of history and the future. No two responses to these images from my students have been the same, which has helped stimulate conversations, encouraging the students to read the image rather than simply look at it.

Years later, when researching environmental social movements, I found myself capturing very different kinds of images. Extinction Rebellion activists at a protest in London in 2022, holding placards explaining the perils of continued extraction and use of fossil fuels (Figures 12.8–12.9) made for rich contrasts – the relationship of the words on the placards with the activists' facial expressions and body language, the juxtaposition between the protesters and the passers-by, between the government buildings in the background and the human mass in the foreground. I was reminded of my college art professor who would make my art class spend hours drawing the details of spaghetti strewn all over the floor, telling us: "Everything you need to know about art is in the relationships between the bits of pasta". As I was taking photographs

Figure 12.5 A journey to where? (courtesy of Peter Sutoris)

of the protests, I thought that everything one needed to know about climate change was, in one form or another, in the visual elements in front of me and the relationships between them. This is also how I have sought to use these images in my teaching: as layered provocations, mental puzzles and entry points into imaginations of alternative futures.

Images that contain elements of ambiguity and which provide rich material for subjective interpretation are, in my experience, most suited to fruitful conversations with students about the environmental multi-crisis. The pedagogical context in which such images are introduced matters, however. Spaces of critical reflection, shared trust, open dialogue and respectful disagreement are a precondition for fruitful conversations stimulated by ambiguous images, but these are not always easy to build in the context of neoliberal, bureaucratised educational environments. Arguably, an effort to stimulate radical imaginations of learners is counter-cultural within such a system, and often amounts to crossing over into the territory of activism-through-education.

Cooking the visual soup

While observing, consuming and reading visual images can significantly influence the imagination of alternative futures, actively participating in creating images – what is known as "visual writing" in the parlance of visual literacy (Kędra, 2018) – can often be an even more powerful way to stimulate the radical imagination.

Figure 12.6 The crumbling graves I (courtesy of Peter Sutoris)

In my research, I have found methods that incorporate visual writing to be the most fruitful ways to explore young people's imaginations of the future. Inspired by David MacDougall's (2018) work in participatory observational filmmaking, I organised several filmmaking workshops with young people in India and South Africa (Sutoris, 2021) in which they made short videos about subjects of their choice (Figure 12.10). Both of these sites were – much like the Marshall Islands – at the frontier of environmental decay; the workshop in India took place in a rehabilitation site for people forcibly moved by the state during the construction of a large dam, while the South African workshop was located in an industrial area with excessive rates of air pollution and a long history of environmental racism. The short films made by children in both locations contained many images of various forms

Figure 12.7 The crumbling graves II (courtesy of Peter Sutoris)

Figure 12.8 Protesting fossil fuels in London I (courtesy of Peter Sutoris)

Figure 12.9 Protesting fossil fuels in London II (courtesy of Peter Sutoris)

Figure 12.10 A participatory videomaking workshop in India (courtesy of Peter Sutoris)

of environmental activism, which guided my research in exploring environmental learning in activist spaces. The young people, too, found the process to be a learning opportunity very different from the word-driven learning environments of their schools. The films led to many conversations in which the young people read the images they themselves produced, interpreting the choices in scenes, locations, camera angles, editing and sound that went into making their films.

In most education systems, young people typically engage in making visual images in their early education experiences, but as they progress, image-generating activities become rare aside from programmes in art, design, architecture and the like. Many students go through their entire secondary and tertiary education without once engaging in visual writing. Yet, participating in cooking the visual soup through intentional, reflective image-making is arguably one of the best ways to improve visual literacy, to find ways to swim in the soup and ultimately to help keep the radical imagination alive in a world that is too often failing to nurture it.

Concluding reflections: curating the visual diet

When thinking about the role of visual images in education in the context of unprecedented (in humanity's history, at least) environmental decay, I find it helpful to remind myself of what Biesta has described as the "subjectification function" of education.

> [It] might perhaps best be understood as the opposite of the socialization function. It is *not* about the insertion of "newcomers" into existing orders, but about ways of being that hint at independence from such orders; ways of being in which the individual is not simply a "specimen" of a more encompassing order.
>
> (Biesta, 2010, p. 356)

How might educators harness the power of visual images to facilitate subjectification rather than socialisation? The possibilities I have explored in this chapter range from mindful and intentional use of images in the learning process (recognition of the visual soup and its power) to equipping students with tools to read images including reflecting on composition, symbolism, contrast, ambiguity, atmosphere and layered meaning (swimming in the visual soup) to facilitating image-making not just in art classes but across different spaces of education (cooking the visual soup).

Given the complexity and unpredictability surrounding the use of visual images in education, it might be tempting to simply opt out and stick to a focus on verbal literacy. But opting out is not an option because simply by the virtue of living in today's world we have been opted in. Irrespective of whether images are used in educational environments or not, learners are immersed in the visual soup through exposure to social media, TV, advertising and countless other channels. Education has a responsibility to aid young people in finding the tools for curating a visual diet conducive not only to mental health but also to realising their potential as imaginative beings capable of transforming, not simply accepting, the future.

Acknowledgements

The author would like to acknowledge Nuith Morales and Gerald Auten whose thinking has shaped some of the ideas explored in this chapter.

Note

1 This use of "Anthropocene" aims to convey the scale of environmental change, the magnitude of the challenge and the need to shift temporal frames of education from "shallow time" (for example: "21st-century skills") towards deep time. It is not meant to suggest that the responsibility for or the impacts of environmental decay are in any way uniform.

References

Anderson, A. (2012). Climate change education for mitigation and adaptation. *Journal of Education for Sustainable Development*, *6*(2), 191–206. https://doi.org/10.1177/0973408212475199

Barker, H. M. (2013). *Bravo for the Marshallese: Regaining control in a post-nuclear, post-colonial world*. Belmont, CA: Wadsworth.

Biesta, G. (2010). On the weakness of education. In D. Kerdeman (Ed.), *Philosophy of education yearbook* (pp. 354–362). Philosophy of Education Society. https://orbilu.uni.lu/bitstream/10993/7817/1/112%20PUB%20PES%202010%20Weakness.pdf

Brown, D. K. (2001). The social sources of educational credentialism: Status cultures, labor markets, and organizations. *Sociology of Education*, *74*, 19–34. https://doi.org/10.2307/2673251

Ehlers, S., Husain, S., Karboul, A., Kwauk, C., Lovett, S., McBride, L., McCabe, M., Steer, L., & Vaduganathan, N. (2022). *Education for climate action: Why education is critical for climate progress*. Education Outcomes Fund. https://www.educationoutcomesfund.org/post/the-power-of-education-for-climate-progress0bvDIeh3jGtrBDBnHx4

Felten, P. (2008). Visual literacy. *Change: The Magazine of Higher Learning*, *40*(6), 60–64. https://doi.org/10.3200/CHNG.40.6.60-64

Holfelder, A.-K. (2019). Towards a sustainable future with education? *Sustainability Science*, *14*(4), 943–952. https://doi.org/10.1007/s11625-019-00682-z

Illich, I. (1971). *Deschooling society*. New York, NY: Harper & Row.

Jovchelovitch, S., Priego-Hernandez, J., & Glăveanu, V. P. (2017). Imagination in children entering culture. In T. Zittoun & V. P. Glăveanu (Eds.), *Handbook of imagination and culture* (pp. 113–135). Oxford: Oxford University Press.

Kędra, J. (2018). What does it mean to be visually literate? Examination of visual literacy definitions in a context of higher education. *Journal of Visual Literacy*, *37*(2), 67–84. https://doi.org/10.1080/1051144X.2018.1492234

Kishore, A. M., & Karmakar, S. (Directors). (2017). *Nostalgia for the future* [Documentary]. Films Division of India.

Litfin, K. (2016). Person/Planet politics: Contemplative pedagogies for a new earth. In S. Nicholson & S. Jinnah (Eds.), *New earth politics: Essays from the anthropocene* (pp. 115–134). Cambridge, MA: The MIT Press.

MacDougall, D. (2018). How children see. *Journal of the Krishnamurti Schools, 22*, 50–55.

Misiaszek, G. W. (2016). Ecopedagogy as an element of citizenship education: The dialectic of global/local spheres of citizenship and critical environmental pedagogies. *International Review of Education, 62*(5), 587–607. https://doi.org/10.1007/s11159-016-9587-0

Ogden, L. A. (2021). *Loss and wonder at the world's end*. Durham, NC: Duke University Press.

Palinkas, L. A., & Wong, M. (2020). Global climate change and mental health. *Current Opinion in Psychology, 32*, 12–16. https://doi.org/10.1016/j.copsyc.2019.06.023

Peña, E., & Dobson, T. M. (2021). The lost years of visual literacy. *Journal of Visual Literacy, 40*(1), 1–14. https://doi.org/10.1080/1051144X.2021.1902043

Prakash, M. S., & Esteva, G. (2009). *Escaping education: Living as learning within grassroots cultures*. New York, Bern, Berlin, Bruxelles, Frankfurt am Main, Oxford, Wien: Peter Lang.

Reid, A. (2019). Climate change education and research: Possibilities and potentials versus problems and perils? *Environmental Education Research, 25*(6), 767–790. https://doi.org/10.1080/13504622.2019.1664075

Smith, M. (2005). Ecological citizenship and ethical responsibility: Arendt, Benjamin and political activism. *Environments: A Journal of Interdisciplinary Studies, 33*(3), 51–64.

Sutoris, P. (2021). Environmental futures through children's eyes: Slow observational participatory videomaking and multi-sited ethnography. *Visual Anthropology Review, 37*(2), 310–332. https://doi.org/10.1111/var.12240

Sutoris, P. (2022). *Educating for the anthropocene: Schooling and activism in the face of slow violence*. Cambridge, MA: The MIT Press.

Sutoris, P. (Director). (2012). *Undiscovered country: A tale of color and development* [Documentary]. Ann Arbor, MI: Alexander Street Press.

Tanguy, M. (2021, May 19). *Imagery can be healthy or harmful, addictive or nutritious* (D. Rudzāte, Interviewer) [Interview]. https://arterritory.com/en/visual_arts/interviews/25561-imagery_can_be_healthy_or_harmful_addictive_or_nutritious/

Visual Diet. (n.d.). http://www.visualdiet.co.uk

Wallace-Wells, D. (2019). *The uninhabitable earth: A story of the future*. London: Penguin.

13 Engaging young people with ecocritical analysis of language in the media

Billy McClune, Karen Kerr and Jennifer Roberts

Speeding towards catastrophe

This headline phrase is taken from an article published in the *New York Times* (NT 2023). It would come as no surprise to many readers that the headline also includes "Climate Change", and the article refers to "global warming" and a planet that is "dangerously overheating". What might be more surprising would be to learn that the term "global warming" is a relatively recent slogan in the climate change debate (CNN, 2019). The earliest news report on this topic can be traced back to 1912 when a short article "Coal Consumption Affecting Climate" announcing a possible link between coal use and our planet's temperature was published in the *Braidwood Dispatch*, a small Aotearoa New Zealand newspaper (The Conversation, 2022). Science-based news is not new. However, the increasing volume of news, the decision makers who decide what is published as "news", the ease of accessing multiple news platforms and the impact that news reporting can have on global audiences all combine to elevate the individual's capacity to engage with news media and to critically evaluate its language, to that of an important life skill.

The impact of human activity on the environment has touched the public consciousness. Environmental issues are topics of discussion in the public square and are explored in different forms of media and are the subject of public debate. They stimulate concerned activists and influential bloggers who are referenced in newsfeeds on individual social media accounts. Expert analysis and ill-informed comments co-exist, as they are posted, shared and "liked" on social media and feedback platforms.

Media reports and public debate influence personal and public decision-making that affects individuals, communities, society and our planet. Pollution in the form of micro-plastics in the food chain is linked to single-use plastics which any of us may have discarded. Greenhouse gases from burning fossil fuels are linked to decisions about energy that we use for our heating and transport needs. These issues generate a range of responses. For some they engender feelings of hopelessness and eco-anxiety; others are stimulated to make more eco-friendly lifestyle choices, and a few become activists for change.

DOI: 10.4324/9781032634524-18

The negative impact of eco-issues on young people is addressed by Hickman (2020) who illustrates the importance of adults listening to the voice of young people and responding truthfully when providing support to help them critically decipher "the truth" and reliable "knowledge" about the world around them (p. 420). When discussing eco-anxiety, Pihkala (2022) highlights the role of knowledge in contending with the journey through ecological grief to ecological awakening.

We will circle back to youth voice later in the chapter but for now, suffice to say, preparing young people to be critically aware citizens, able to cope with the ecological dilemmas arising from issues they encounter in news and social media, may seem as effecting a duty of care.

Developing ecoliteracy

In their seminal work *Beyond 2000: Science Education for the Future* Miller and Osborne (1998) address the question of what it means to be science literate. They suggest a key indicator of this would be the individual's capacity to read and respond critically to news reports with a science component. This vision for literacy in a science context involves the mastery of essential subject knowledge and technical vocabulary together with an appreciation of how science works. These science knowledge elements sit alongside the promotion of literacy skills. In addition to reading and comprehension, higher-order skills of inference, interpretation and analysis are needed. So, to borrow from Miller and Osborne's (1998) observations on what it means to be science literate we might suggest that in this context, to be "eco-literate", an individual should be able to access and make reasoned responses to sources of "news" about environmental issues that impact them personally, their communities and society.

An established model for developing science literacy in the context of science reported in the media was initially developed in the "Science Newswise" project. This highlights four core strands in an educational framework. They are subject knowledge, literacy skills, media awareness and critical habits of mind (Jarman & McClune, 2007). In addition to knowledge and skills, it is considered that the critical media consumer needs to have a degree of "media awareness" if they are to adequately interrogate the media source and "an inquiring habit of mind" that characterises the active, rather than passive, consumer who exhibits curiosity, a reasonable degree of scepticism and a desire to know both sides of an argument (McClune & Jarman, 2011). This science literacy model might equally provide a foundation for the development of ecoliteracy in the classroom.

Accepting the cross-disciplinary challenge

The combination of knowledge, skills and attitudes that characterise this view of ecoliteracy fall outside the scope of any single subject area as normally defined in the school curriculum. Science-based news reports are cutting-edge rather than textbook science; they exemplify non-fiction writing. The factual content of media sources may appeal to science teachers while language arts teachers looking for non-fiction texts may bypass the unfamiliar subject content of science in favour of more accessible topic material. However, in the context of cross-disciplinary collaboration, this type of text may warrant the closer attention of both science and language arts teachers.

If the intentions of the model for critical engagement with environmental issues in the news media are to be realised, then it could be argued that interdisciplinary collaboration is an inevitable outcome. Insights into the development of a foundational framework, and pedagogical approaches that emerged from collaborative work of media experts, language arts specialists and teachers of science are offered by Jarman & McClune who note that:

> When teachers introduce science related news items into their classrooms they are, inescapably, setting out on a multidisciplinary path ... drawing on knowledge and skill including but also beyond that normally identified with learning in science. As they attend to language, as they distinguish "fact" from opinion, as they explore points of view, they are using abilities more commonly associated with English and media studies. As they weigh rights and responsibilities, consider ethical and moral dimensions of an issue they are using abilities more commonly associated with humanities and religious studies.
>
> (2007, p.162)

Indeed, the strategies and action required to develop new and existing solutions for the nature crisis will take a strong multidisciplinary approach and is an important approach to model with young people, as we develop their science and ecoliteracy.

Exploring traditional News media

If news reports are to be both the means to and the goal of science or ecoliteracy education, then the nature of news in general and science-based news reports in particular needs to be acknowledged. Science-based issues in the news are often complex. The topics are diverse, and consequently, the scope of foundational knowledge and the range of technical vocabulary needed for public engagement can be sizeable.

Making news is a process of purposeful selection. "Newsworthiness" is an overarching term that helps explain why some events are publicised and others are not. Selection is at the core of all media reporting and also news consumption. With a search term and the click of a mouse, the reader can pick their news. This selection process, with other factors, combine to act as filters in how we allow second-hand information to influence our view of the world around us. The complexity of the topic, the selectiveness of the journalist and editors, and the randomness of the search present those wishing to engage learners with media-based reports about environmental issues with a challenge. How do you help young people in a classroom context negotiate the media landscape and distinguish between multiple influences, evaluate the credibility of the report and learn to make personal and reasoned responses?

Embracing modern media

Social media use has grown exponentially in the last decade, from an estimated 43% of the population with an active mobile social account in 2014 (We Are Social, 2015) to 84% of the total population of the UK in 2023 (We Are Social, 2023). As children and young people get to grips with the media in all its various forms, they will need to develop appropriate interpretative skills. Studies suggest that more young people than not are interested in news. Just under six in ten 12- to 15-year-olds claim to be either "very" or "quite" interested in news (Ofcom, 2022). Family talk and television are reported to be the most common ways that teens find out about news. However, new media sources are attracting a following of young people. Table 13.1 illustrates the changing nature of media sources that people use to follow the news.

Studies have shown that different age groups consume news differently. Young people associate news awareness with the context of family talk and television. However, they report that social media are their preferred platforms to follow news stories. A recent Ofcom survey on news consumption concluded that younger age groups are much more likely to use the internet and social media for news, whereas their older counterparts favour print, media, radio and television. In a survey of young people aged 12 to 15 almost one in three reported using Instagram, TikTok and YouTube to follow news stories, noting an upward trend in the use of Instagram and TikTok. Over a five-year period to 2022, the use of more traditional news sources such as television declined noticeably among this age group. However, more conventional broadcasting platforms still attracted one in four of those surveyed. Radio, print media and news websites remained largely unchanged at a low level of less than one in ten users (Ofcom, 2022). It is perhaps fair to conclude that among teens, social media are overtaking traditional news channels.

Table 13.1 A summary of news sources 2010–2022

Leading news outlets (Jarman & McClune, 2010)	*Preferred sources in rank order (Ofcom, 2022) (Age group 12–15)*
• Newspapers • Television (scheduled news programmes) • Radio • Magazines • Curated articles	• Instagram • Tik Tok • YouTube • Television • Facebook (incl. Messenger) • Snapchat • WhatsApp • Twitter

How do young people interact with media?

Key debates on how young people interact with modern media often consider factors other than news content, for example, concentration and how long users interact with certain forms of media. Whilst all age groups spend more time overall engaging with online news, Thurman and Fletcher (2019) found that for younger readers (like for middle-aged and older consumers), newspapers' print editions provide an experience they invest time in, compared with how they "snack" on, and scan, news online. As well as increased consumption overall, there is a trend for young people to spend less time per story, thus creating a larger number of momentary encounters where large volumes of content are presented during a single scrolling session.

Young people are less likely to engage with long-form videos. One in four of the most liked TikTok videos falls within the optimal recommended video length of 21 to 34 seconds (WIRED, 2022). All of this indicates that shorter videos perform best with the TikTok community.

What skills and capabilities do young people need?

Young people are exposed to a wider variety of news from an increased range of opinions across multiple platforms and formats. To engage critically, people need to be able to spot sponsored posts, identify source material and understand the nuances of a given app. In addition, consumers need to be aware of how algorithms that are used to manage content make it more difficult to find material viewed previously, and this limits the option to revisit content. Awareness of the structure and function of social media provides a foundation for critical engagement.

Ofcom (2022) found that most 16–24-year-olds claim to access news on social media. However, while significant proportions of young people use social media to follow news stories, they remain sceptical about its trustworthiness. Also, almost four in five 12- to 15-year-olds said the news they heard from family was either "always" or "mostly" accurate. Their confidence in broadcast media was similar though less fulsome. Fewer believe news stories from friends or social media are completely accurate.

The skills that enable a consumer to translate scepticism into discernment are needed in this modern media environment where trust is not unconditional, but media sources and peer involvement are persuasive. The skills and abilities needed to understand and interact with media are expanding to include characteristics unique to media platforms that influence news. This may involve some reimaging of media awareness and discerning habits of mind. Table 13.2 summarises the shifting skills burden for critical engagement.

The way young people receive and interact with news is changing. Young people are aware of the different news forms available and are increasingly turning to social media for information. The algorithms which drive social media, and the way short-form news is produced on many social media apps,

Table 13.2 Media-focused critical engagement skills and capabilities 2010–2022

Critical engagement skills and capabilities	
2010	*2022*
Language analysis skills Awareness of aesthetics impact Appreciation of statistical information Identification of news sources Analysis of article tone Consideration of author credibility	Language analysis skills Awareness of aesthetics impact Appreciation of statistical information Identification of news sources Analysis of article tone Consideration of author credibility Identification of paid content and sponsorship Consideration of the origin of news item Awareness of algorithm impact Quick decision making about transient content

can hinder understanding and can make it more difficult for young people to know what and who to trust.

While it is important to acknowledge the wider skills framework needed for this type of work, particularly given the dynamic context of media consumption and skills required for young people, it is important not to lose sight of the science element and, in turn the "eco-science" elements. As critical engagement with science in the media requires scientific literacy, engagement with "eco-science" in modern media requires critical ecoliteracy.

From scientific literacy to critical ecoliteracy

In their review article, McBride, Brewer, Berkowitz & Borrie (2013) explore the essential elements of ecoliteracy and the allied constructs of ecological and environmental literacy. They note efforts being made to define and delineate key terms in this field but highlight the confusion caused by indiscriminate use of terms. Much of this discussion is reminiscent of an earlier debate around an older but similarly contested term: "science literacy". That term accommodates both the mastery of essential subject knowledge and technical vocabulary alongside the promotion of literacy skills needed for understanding and application of knowledge. Such an understanding would also benefit the ecoliteracy debate and what constitutes critical ecoliteracy.

Critical ecoliteracy

Whilst ecoliteracy is an accepted and well-known term, it is a broad umbrella term and several scholars have examined the importance of the development of an accepted definition and exploration of how critical ecoliteracy develops. Kahn's (2009) outworking of the "critical ecoliteracy" construct is thought-provoking and apposite in the context of an environmental emergency that has global implications. From his perspective, systems and structures, including media that emerge from cultures and society, ensure that the environmental issues are variously portrayed in all aspects of media in ways that either promote "ecological sustainability" or lead to pathways of "unsustainability and extinction". For example, a view of newsworthiness that promotes fast fashion brands by influencers driving consumerism via "sponsored" recommendations could be seen to promote unsustainability.

Critical ecoliteracy brings together both personal and societal elements of responsibility. It addresses personal decision-making along with the responsibility to hold society to account. The focus of critical ecoliteracy includes personal responsibility; it is not just that which sits beyond or above the individual. Consequently, Kahn (2009) raises the importance of developing an awareness of a person's own critical literacy and an understanding of how this interacts with an ability to affect the future. This is addressed in part in the development of media awareness noted previously by Jarman and McClune (2010) as a component of science literacy. In this context we might introduce the term *eco-curiosity* whereby as an "active" consumer of media, the development of an inquiring habit of mind might be described in terms of a healthy scepticism and a desire to know both sides of an argument.

This is well illustrated in the response to *blame narratives* around the environmental issue of climate change. Rudiak-Gould (2014) summarised these as "industrial blame" which focuses on industry and developed countries and "universal blame" that attributes responsibility to all human beings (p. 366). This latter term however is nuanced, as Crate and Nuttall (2009) note: the people who are most susceptible to the impact of climate change contribute the least.

It could be argued that critical ecoliteracy requires the development of knowledge, literacy skills, media awareness and critical habits of mind – the four core strands of scientific literacy identified by Jarman and McClune (2007). In addition, the more recently cited "qualities" required to be critically eco-literate sit with, and give more breadth and depth to, these core strands. Building on Kahn's (2009) exploration of critical ecoliteracy, Turner (2011) suggested a list of eight qualities that critically eco-literate individuals might display. These included "imagination" (to think of different ways to do things and live our lives), "ethical consciousness" (questioning right and wrong; thinking through our responsibilities to others) and "critical language awareness" (including language in the media) (p. 45). Each of these has application as we think about responding to media reporting of ecological issues. Of these, critical language awareness has unique resonance.

Turner (2011) defines Critical Language Awareness, in the context of critical ecoliteracy, as "an examination of how language is used, by whom, and why, and how these uses of language affect our relationship to the more-than-human-world" (p. 46). Turner goes further to specify that (young) people must develop a "critical awareness of the workings of language, communication, and media" and, very importantly, how these influence "our views of the world" (p. 56). Turner calls on education to provide (young) people with:

> the conceptual tools to critically examine the influence of language and communication on their thinking and actions, and to understand and analyse how language is used, to what effects, and for what (and whose) purposes.
>
> (2011, p. 57)

Core ideas: from theory to research-led practice

The argument advanced in this chapter is that environmental and ecological literacies are strands within the wider construct of science literacy. The development path to ecoliteracy closely mirrors that to science literacy. With a focus on language in the media, this chapter highlights strategies and pedagogical practices that are intended to develop the skills and habits of mind that will support a critical approach to media-based texts. These along with a willingness to embrace interdisciplinary working are integral to a framework for modern-day eco-pedagogy. When framing an approach to eco-pedagogy it is important to consider the context in which such material will be delivered. One of the common issues cited by teachers relates to difficulties with relating newsworthy ecoliteracy issues to existing curricula content. The following examples and approaches illustrate ways in which practitioners can use a critical ecoliteracy lens to help develop awareness of language in the media in the classroom.

Promoting critical reading of science-based issues presented as news

Below is the headline of a news report published in the online version of Advanced Science News (Figure 13.1). Similar accounts of this and other related research can be found in other online outputs of print media and web-based news outlets. The news article describes the outcome of research from a reputable university, by named scientists and reports on a paper published in a peer-reviewed academic journal. It is an example of science-based news that is readily found using an internet search engine.

ADVANCED SCIENCE NEWS

TURNING PLASTIC WASTE INTO LOW-COST HYDROGEN FUEL

By Victoria Corless September 18, 2023

A flash heating technique breaks down plastic waste and converts it to pure hydrogen and graphene with significantly less emissions and at a low cost.

Figure 13.1 Example of the headline and topic sentence from a news article

The article is précised by the single opening sentence above. In its complete form, this is linked to an evocative image of discarded plastic waste. The topic sentence concisely describes the main point of the story and introduces key science vocabulary which in the classroom context, may need further explanation, probably from a science teacher. The article goes on to explain the process of *flash heating* which passes an electric current through the plastic. The authors quote one of the researchers who describes the context prior to the start of their project. He explains "We know how to recycle plastic; the problem lies in the fact that recycling is expensive and there is no economic initiative incentive"; and again "Producing hydrogen fuel without also producing carbon dioxide (a greenhouse gas) is possible but more expensive than other methods". This puts in context the significance of the claim, reported in the article, that the researchers have developed an economically viable process to convert plastic waste into high-value products.

The basis of their research (a novel process that converts waste plastics into clean hydrogen and graphene) is then further explained. The article notes that "Plastic is mainly composed of atoms of carbon and hydrogen". In this process of flash heating scientists "reorganise the chemical bonds in plastic to produce graphene (a valuable form of carbon) and hydrogen (a clean fuel)". The whole process is free of polluting by-products, and the value of the graphene itself outweighs the cost of the process.

Researchers align their work with the EarthShot Challenge (2023) which seeks to mobilise resources and creativity to achieve ambitious environmental goals. However, they also acknowledge that "scale-up" is still an issue. As hydrogen is a flammable gas, its safe capture and purification requires some careful planning and engineering (*Advanced Science News*, 2023).

Strategies and resources in text

Previous studies, in the context of print and online versions of news, have identified a range of resources that are effective in promoting critical reading of science reported as news. These have been summarised in a three-stage framework to promote critical reading. The first stage focuses on reading and comprehension. The aim is to encourage pupils to examine the text by reading it closely and using key questions to test the credibility of the text. The second stage focuses on vocabulary and language. The aim is to evaluate the text by considering the impact of language. Finally, the third stage focuses on reasoning and argument. The aim is to encourage a reasoned response to the text (McClune & Alexander, 2015).

Examining the text

Pupils should be able to test the credibility of the text using some key questions. One activity that has had a positive impact in the classroom is the "News Bug". Each of the legs of the bug represents a critical question that can be used to interrogate the text. Pupils can assemble or disassemble the bug one question at a time. A news bug with insufficient legs is not able to stand up and might suggest that the news report lacks credibility in some key areas. It is important to note that if a news report is considered to be inadequate it does not necessarily follow that the science being reported was flawed.

Applying this strategy to the news report on scientists' claim to turn plastic waste into hydrogen fuel would confirm that the report passes the credibility test. The report provides information about the author and source of the report, the names of the scientists and the institution where the work was carried out and the peer-reviewed academic journal where their findings were published. In addition, there is theoretical and practical information. To support the claims, however, no observations or opinions from other scientists are included.

Evaluating the text

Pupils should assess the language used in the text against some key criteria. Identifying the sources used, taking note of limiting clauses, being alert to persuasive or emotive language and distinguishing between facts and opinions are all useful skills for the critical reader to acquire. Taken together, these skills can be effectively addressed in the classroom. Using a "certainty meter" in an activity means that selected words and phrases from the text can be examined, identified as fact or opinion and rated according to their perceived degree of certainty. If this strategy was applied to the article on plastic waste above, pupils might note the quotations from the researchers and their level of certainty about the scientific process of flash heating or the facts about the chemical composition of plastic. However, in relation to scaling up the work from the laboratory to industrial production of the fuel, the article is less certain.

Provoking a reader response

Critical reading is an active process, so in the classroom context pupils should be encouraged to acknowledge and examine the responses that their reading of the article provokes. What questions does it raise? A response may include assessing an argument that is being put forward, or seeking additional information by looking for other news reports on the same topic. In the plastic waste news report pupils should decide if the subject of the article is important. Pupils might consider if the claims have significance on a personal, local or on a global scale. They could be encouraged to list the claims and ask what supportive evidence is provided. Finally, pupils might be encouraged to summarise the article in fewer than 50 words. Alternatively, given the exemplar summary to read (quoted below), pupils could be invited to post their comments on the extract before responding to the comments from others:

> The problem of discarded single use plastic has global impact. The ability to convert discarded plastic into an environmentally friendly fuel would appear to offer a way to manage a growing waste problem and help to solve the problem caused by burning fossil fuel to generate energy.

Strategies and resources for news on social media formats

Embracing social media

The news report in Figure 13.1 was accessed in an online version. The same and similar research is also referenced in other social media platforms. A mix of traditional media activities and more modern media sources is important; it reflects the way young people receive news naturally and helps them develop skills and to understand the language of the social platforms they are widely using. Pupils could be encouraged to make their own audio online report in a particular social media style. The following four activities are sample classroom activities which allow pupils to reflect and work on new media and traditional media types.

One-minute news

This activity asks pupils to think about what the most important information is to include in news reports (either written or oral). It is linked with curriculum elements around reading and understanding non-fiction texts, digital media and speaking and listening. Pupils are given a news report on a science-based issue such as the article referenced in Figure 13.1 and are asked to create a one-minute presentation based on the issue. Students are then asked to create an oral report on the same theme but only 30 seconds long. They then reduce the report again, so that it is only 15 seconds long. This activity helps students to think about long-form and short-form reporting. Follow-up reflections ask them to consider how they prioritised information and how they chose what to leave in.

Headlines and hashtags

This activity asks pupils to think critically about author/creator's intent and how the words chosen can impact our understanding. It links with curriculum elements around reading non-fiction texts and critical literacy. Pupils look at a variety of headlines and news reports (both print and digital). They also look at social media posts and the hashtags used before discussing how the messages have been twisted and in what way the words chosen have impacted the consumers' understanding of the story. This activity concludes with a question and an invitation. How can we look critically at headlines and hashtags – what are they designed to do? Using some examples of text or images taken from ecological issues in the news, students can be invited to make their own headlines or hashtags.

Picture this

Images tell a story, and this activity looks at visual media and the impact a photograph can have. It is linked with the digital media elements of the curriculum. There are two related tasks in this activity. Firstly, pupils analyse the photographs and "match" these to the headline that they think would attract the reader to the story, or to issues they think the picture is exploring. Exemplar articles such as that in Figure 13.1 could be used as stimulus material. Teachers could develop this task by providing pupils with headlines from environmental issues in the news and then by inviting them to locate a suitable image to promote the news story.

#AD

As media have changed and sponsored content and social media influencers have grown in popularity, this task asks pupils to identify sponsored content and compare it to traditional advertisements. Pupils look at traditional written and video advertisements and watch some sponsored content from YouTube/TikTok and Instagram. They can then compare and contrast the forms and begin to think critically about how consumers of media are influenced by each type of content.

Interdisciplinary collaboration

In the classroom: the newswise project

By exploring the cross-disciplinary nature of science-based news reporting, the "Newswise" Project initiated collaborative work in schools involving teachers of Science and teachers of English and developed a pedagogical framework and supporting resources for pupils at the primary and secondary level. The outcomes recognised the different contributions that each subject made to a young person's capability to read, understand, and make a reasoned response to news reports for the science component (Jarman & McClune, 2011).

In the context of science-based media education there are both pedagogical and epistemological benefits to interdisciplinary collaboration. However, the challenges for teachers in trying to integrate this type of collaborative work in the classroom should not be underestimated. There are time pressures around the necessary subject content within the curriculum and operational obstacles due to timetabling. In addition, personal preferences and traditional working practices can hinder change (Alexander, Jarman, McClune & Walsh, 2008). Communication of subject knowledge and scientific principles or ways of knowing are within the domain of science. The development of literacy skills and media education are more commonly, though not exclusively, the domain of language and media arts education.

In teacher education

It is perhaps not surprising that the subject background of teachers influences their critical appraisal of media reports. A study involving both science and language arts pre-service teachers demonstrated the different but complementary perspectives that each of these groups has in relation to news reports for the science component. This study identified eight indicators of critical engagement with science-based media reports. Half of these, characterised as language elements, were mostly highlighted by language arts teachers. Science teachers mainly highlighted elements characterised as the nature of science. The report concluded that interdisciplinary collaboration is essential if pupils are to be introduced to a comprehensive experience of a critical approach to science-based media (McClune & Alexander, 2015).

Similarly, a recent application of an existing co-teaching model (Kerr, 2020a; Murphy & Beggs, 2010) was piloted with English and Science pre-service teachers. The pilot project saw a Science specialist and an English expert bringing separate expertise to co-plan, co-teach and co-evaluate a series of lessons titled "Environmental and Sustainability Education for All". The pre-service teachers engaged in workshops designed to focus on the theory and practice of using a co-teaching approach and capacity building around teaching about the environment.

Looking forward: youth voices and social media as critical ecoliteracy

The impact of youth voice and activism has been a hot topic in the media (including social media) in more recent times. Aside from the very influential activism of Greta Thunberg and the School Strike for Climate, there are several other examples across the world and in the UK, to include Teach the Future, Our Bright Future Youth Forums and Ulster Wildlife's Youth Forum.

Teach the Future is a youth-led campaign with the key aim of repurposing the curricula (across the UK) around the climate emergency and ecological crisis. Ulster Wildlife's Youth Forum has also published a Nature Manifesto (Ulster Wildlife, 2022) with key "asks" for the next Programme for Government centred around "For Nature", "For Climate" and "For Young People". In a similar vein to the call from Teach the Future, this Youth Forum wishes to see specific teaching on nature within school education.

Whilst the young people in these organisations are already engaged in protecting our environment and taking action, the importance of engaging their less eco-literate peers is acknowledged. An evaluation of Ulster Wildlife's Grassroots Challenge project (Kerr, 2020b) evidenced positive outcomes as a result of being involved. It was set up to offer essential skills, knowledge and training to take action for wildlife and was designed for young people aged 16–24. During in-depth focus groups with the young people who were involved in the Grassroots Challenge, it became very clear that they are concerned about the lack of awareness of environmental issues among their peers, that there is still "a stigma" attached to being involved in environmental issues and a very real lack of opportunity and promotion of existing opportunities to take part in activities to make a difference for nature. In one focus group, there was an emphasis placed on the types of people (influencers) on social media and the things they push on young people (for example makeup) which are so much at odds with what they felt should be trending and influencing young people – the state of nature and what they can do to support it.

This work highlights an opportunity missed – social media is highly influential in the lives of our young people and, as suggested by young people themselves, the perfect vehicle to influence a climate call to action. Critical ecoliteracy across the curriculum delivered in an interdisciplinary way, as highlighted in our media-focused, eco-pedagogy examples, offers an opportunity to address this call from our young people.

Conclusion

Media reports on eco-based issues regularly appear in news outlets and are widely discussed on social media platforms, so it is important that young people are equipped to engage critically in both community debate and the personal decision-making that this public argument engenders. Eco-critical analytical skills help to equip citizens to appreciate how media presentation can shape their view of issues, seek out a range of sources to understand different perspectives, engage confidently with issues that are important to them and make reasoned responses. Supporting young people to engage critically with sources of knowledge will empower them to address ecological issues and in so doing may also support their emotional journey.

An appreciation of the range of skills and knowledge that underpin a comprehensive view of eco-critical analysis should encourage teachers and curriculum planners to value and advocate for cross-disciplinary and collaborative approaches which provide learners with a more coherent understanding of the complex issues involved in becoming ecoliterate. Models and strategies for developing pupils' critical appreciation of science reported in the media are available, and these provide starting points for a pedagogical framework, using a blend of language arts and science methodologies, aimed at promoting ecoliteracy. However, while nature science is a foundational epistemology, applying this to the specific subject matter of ecology and taking account of the ever-changing landscape of broadcast, print and social media platforms will provide teachers with an ongoing challenge.

References

Advanced Science News. (2023). Turning plastic waste into low-cost hydrogen fuels. By Victoria Corless. Retrieved October 7, 2023, from https://www.advancedsciencenews.com/turning-plastic-waste-into-low-cost-hydrogen-fuels/

Alexander, J., Jarman, R., McClune, B., & Walsh, P. (2008). From rhetoric to reality: Advancing literacy by cross-curricular means. *The Curriculum Journal*, *19*(1), 23–35. https://doi.org/10.1080/09585170801903225

CNN. (2019). Is it climate change or global warming? How science and a secret memo shaped the answer. By Jen Christensen. Retrieved October 13, 2023, from https://edition.cnn.com/2019/03/02/world/global-warming-climate-change-language-scn/index.html#:~:text=Global%20warming%3A%20A%20beginning,of%20a%20Pronounced%20Global%20Warming%3F%E2%80%9D

Crate, S. A., & Nuttall, M. (2009). Introduction: Anthropology and climate change. In S. Crate & M. Nuttall (Eds.), *Anthropology and climate change: From encounters to actions* (pp. 9–36). Walnut Creek, CA: Left Coast Press.

Earthshot Challenge. (2023). Retrieved October 16, 2023, from https://earthshotprize.org/our-vision-mission/

Hickman, C. (2020). We need to (find a way to) talk about . . . Eco-anxiety. *Journal of Social Work Practice*, *24*(4), 411–424. https://doi.org/10.1080/02650533.2020.1844166

Jarman, R., & McClune, B. (2007). *Developing scientific literacy: Using news media in the classroom*. Maidenhead: McGaw Hill/Open University Press.

Jarman, R., & McClune, B. (2010). Developing students' ability to engage critically with science in the news: Identifying elements of media awareness dimension. *The Curriculum Journal*, *21*(1), 47–64. https://doi.org/10.1080/09585170903558380

Jarman, R., & McClune, B. (2011). *Science newswise 2*. Hatfield: Association for Science Education.

Kahn, R. (2009). Towards ecopedagogy: Weaving a broad-based pedagogy of liberation for animals, nature, and the oppressed people of the earth. In A. Darder, M. P. Baltodano, & R. Torres (Eds.), *The critical pedagogy reader* (pp. 522–540). New York, NY: Routledge.

Kerr, K. (2020a). Teacher development through coteaching outdoor science and environmental education across the elementary-middle school transition. *The Journal of Environmental Education*, *51*(1), 29–43. https://doi.org/10.1080/00958964.2019.1604482

Kerr, K. (2020b). *Report for Ulster wildlife: An evaluation of the Grassroots Challenge Project 2019–20*. Belfast: School of Social Sciences, Education and Social Work, Queen's University Belfast.

McBride, B. B., Brewer, C. A., Berkowitz, A. R., & Borrie, W. T. (2013). Environmental literacy, ecological literacy, ecoliteracy: What do we mean and how did we get here? *Ecosphere*, *4*(5), 1–20. https://doi.org/10.1890/ES13-00075.1

McClune, B., & Alexander, J. (2015). Learning to read with a critical eye: Cultivating discerning readers of media reports with a science component. *School Science Review*, *97*(359), 21–28.

McClune, B., & Jarman, R. (2011). From aspiration to action: A learning intentions model to promote critical engagement with science in the print-based media. *Research in Science Education*, *41*(5), 691–710. https://doi.org/10.1007/s11165-010-9186-1

Miller, R., & Osborne, J. (Eds.). (1998). *Beyond 2000: Science education for the future*. London: Kings College.

Murphy, C., & Beggs, J. (2010). A five-year systematic study of coteaching science in 120 primary schools: Coteaching in continuing professional development. In C. Murphy & K. Scantlebury (Eds.), *Coteaching in international contexts research and practice* (pp. 11–34). Dordrecht: Springer.

NewYork Times. (2023). *Climate change is speeding toward catastrophe.* By Brad Plumber. Retrieved October 13, 2023, from https://www.nytimes.com/2023/03/20/climate/global-warming-ipcc-earth.html

Ofcom. (2022). *News consumption in the UK: 2022.* https://www.ofcom.org.uk/_data/assets/pdf_file/0027/241947/News-Consumption-in-the-UK-2022-reportpdf

Pihkala, P. (2022). The process of eco-anxiety and ecological grief: A narrative review and a new proposal. *Sustainability, 14*(16628). https://doi.org/10.3390/su142416628

Rudiak-Gould, P. (2014). Climate change and accusation: Global warming and local blame in a small island state. *Current Anthropology, 5*(4), 365–386. https://doi.org/10.1086/676969

The Conversation. (2022). *For 110 years climate change has been in the news. Are we finally ready to listen?* By Linder Ashcroft. Retrieved October 13, 2023, from https://theconversation.com/for-110-years-climate-change-has-been-in-the-news-are-we-finally-ready-to-listen-188646

Thurman, N., & Fletcher, R. (2019). Has digital distribution rejuvenated readership? *Journalism Studies, 20*(4), 542–562. https://doi.org/10.1080/1461670X.2017.1397532

Turner, R. (2011). Critical ecoliteracy: An interdisciplinary secondary and post secondary humanities curriculum to cultivate environmental consciousness. (Doctoral dissertation, University of Maryland, Baltimore County, MD). ProQuest Dissertations Publishing.

Ulster Wildlife's Youth Forum. (2022). *Youth manifesto for nature and the environment.* Ulster Wildlife.

We Are Social. (2015). *Digital 2015 the United Kingdom.* Retrieved July 26, 2023, from https://www.slideshare.net/DataReportal/digital-2015-united-kingdom-january-2015

We Are Social. (2023). *Digital 2023 the United Kingdom.* Retrieved July 26, 2023, from https://datareportal.com/reports/digital-2023-united-kingdom

WIRED. (2022). *TikTok wants longer videos—Whether you like it or not.* Retrieved October 26, 2023, from https://www.wired.co.uk/article/tiktok-wants-longer-videos-like-not?utm_source=twitter&utm_medium=social&utm_campaign=onsite-share&utm_brand=wired-uk&utm_social-type=earned

Conclusion

Ian Davies

Introduction

McGuinn and Naylor provide a clear and necessary focal point that has motivated the authors of the chapters of this book. In their compelling phrase: "This book makes a case for the role which language arts might play in environmental education". This aim has been magnificently achieved in the first main section opening up the debate with reflections and insights on the environment and social inequities, ecocritical reading, the humanisation of science and the need to encourage resilience in the face of environmental challenges. The subsequent chapters that explore and provide ways forward for what the language arts can offer demonstrate expert argumentation around types of arts (including language, drama and visual images), the role of place and the perspectives that are possible and desirable, and give hugely helpful discussions around the outcomes of educational programmes (for example, enhanced understanding, the liberation of voice and practice of advocacy and the taking of practical action with the necessary educational reflective stance on what has been achieved).

In this final chapter, I will not repeat what has been already achieved by the team of authors from so many different parts of the world. Rather, I will use their ideas as a springboard in an ongoing discussion, respecting their work, inferring from it what I think are some ideas and issues that are key to the achievement of the central purpose of the book. Following some brief contextual remarks about environmental education, I will explore the characterisation of the environment through a metaphorical approach, discuss ways to outline one possible overarching educational framework and, prior to a brief conclusion, highlight a very few of the ways in which interventions could be made to ensure the realisation of a valuable form of environmental education that is informed, supported and realised by and through language arts.

Context

It is necessary to be clear about the meaning of key terms. My view of the environment is that it is the places where we live, what we do in those places and who we are. My sense of the arts, precisely the language arts, is that they

DOI: 10.4324/9781032634524-19

allow us to gain insight into our humanity, to express what it means to be human and by doing so to generate more deeply meaningful characterisations of that humanity. The editors and authors of this book have taken the ambitiously bold step of bringing the environment together with language arts. In so doing, they are concerned with Education for Sustainable Development (ESD). Writing four years after the launch of the United Nation's *Decade of ESD*, Arjen Wals notes the following key areas of "common ground" associated with ESD:

- A transformative and reflective process that seeks to integrate values and perceptions of sustainability into not only education systems but one's everyday personal and professional life.
- A means of empowering people with new knowledge and skills to help resolve common issues that challenge global society's collective life now and in the future.
- A holistic approach to achieve economic and social justice and respect for all life.
- A means to improve the quality of basic education, to reorient existing educational programmes and to raise awareness. (Wals, 2009, p. 26)

This task means that several almost overwhelmingly complex and fundamental challenges must be faced. There is substantial evidence of an elemental planetary crisis. Kemp (2023) draws attention to alarming data that includes a vast amount of material including that from NASA's Goddard Institute for Space Studies (GISS) 1880–2022 about global warming. There is now so much data and so many very high-profile initiatives (for example: United Nations, 2023) with summits and statements (for example: Carnegie Europe, 2021) that it is simply obvious that we are experiencing an extremely serious crisis.

And yet, almost unbelievably, there are frequent abdications of responsibility by governments and many others. There are literary points of recent interest from the global north at various points (for example: in the work of Thoreau) and in key events in decolonialisation (for example: Gandhi in the 20th century). But perhaps the most significant profile for the environment may be seen principally in ancient societies (Fredericksen, 2019) including those from the east (McAuley, 2023) and, of course, in the ideas and practices that are embedded in Indigenous cultures. Although there are exceptions, the environment gains traction in the global north only from the 1960s as a part of various social movements (and even then, in the context of high-profile panics not to save the environment but about the widely predicted exhaustion of oil reserves). From the 1990s onwards, Earth Summits and COP meetings gained rhetorical attention and, simultaneously, were disregarded. At times, this neglect and opposition is a wilful rejection of evidence as part of a particular political stance. Trump's approach to the environment during his time in office as US President (see Lewis, 2018) is illustrative here.

In part, this neglect is due to the contested nature of the field. Ideas about the environment struggle to achieve consensus. At points, the refusal to safeguard the environment is portrayed as responsible action, indicating that, for some, there is agreement over means but not ends. Recycling, for example, may be seen by some as the way to achieve a form of sustainability with the green agenda incorporated into mainstream neo-liberal political positions. The changing nature of what we understand by the environment and what is needed may be seen in the different approaches to the relationship between literature and the environment. A range of ideas and perspectives emerge at different points. Buell, Heise and Thornber (2011) show the different aspects of our relationship with the environment and literature, involving varying emphases on place, science, gender, postcolonialism, indigeneity and animals. The very titles of the field (which I am somewhat glossing over but which with greater space would receive more attention) reveal different interpretations, preferences and emphases. "Sustainability" for some is not as useful as "environmental" while others prefer "ecocentrism" (and many other options exist). According to Huckle (2012), the key divide is "between reformists and radicals: between those seeking the greening of capitalism and those seeking the greening of socialism" (p. 363). There are those who prefer what they regard as market efficiency with sustainable consumption managed by national experts rather than an appreciation and recognition of sufficiency achieved in part through regulation including localism.

As a result, in part, of these debates about the environment itself, too often there is a neglect of education about, through and for the environment. Of course, there is rhetorical commitment. Who would be willing to declare an antipathy towards the environment and who would not expect education to play a part in its protection and enhancement? The nature tables in classrooms in elementary and primary schools, the idealism of young people who become involved in local, national and global schemes and the obviously good work of so many in charities and NGOs as well as in research and teaching programmes need to be recognised. Many mainstream groups now assert the need to recognise the link between the arts and understanding and acting in contemporary society (for example: NECE, 2023a). But environmental education is relatively new; it is not regarded by many as a "proper" discipline or standalone subject leading to high-status, school-age, externally governed examinations, and is often a phantom curriculum, talked about but never seen. It suffers from the fatal combination of being low status and high risk. Teachers are already subject to official guidance about political impartiality (for example: in England, see Department for Education, 2022) and are not unaware of the professional risks they may be subject to if they say or show the "wrong" thing in the classroom (MacLeod, 2007). They are under significant attack from those whose neo-liberal ideology is presented, falsely, as being apolitical.

In the context of neglect, opposition and a lack of consensus over key ideas and practices, this book plays a vital role. It is important to challenge complacency and oppose those who have not yet realised the severity of the current crisis; to clarify the meanings of key ideas in ways that demonstrate the positive possibilities that lie before us; and to describe and discuss the very practical ways forward that need to be considered. Academically, professionally and in educational encounters in all contexts, this book sheds light on complex matters and illuminates the roads that may be taken.

Metaphorical characterisations of the environment with illustrations from literature and linkages with political thought

If we are to learn and teach about, through and for the environment, we have to be clear about our understanding of it. In this section, I use the literary device of the metaphor to illuminate the ways in which we portray the environment, revealing not just what we express but what we think. As Gibbs (2011) suggests, "metaphor is not just an aspect of language, but a fundamental part of human thought" (2011, p. 529). Drawing from conceptual metaphor theory, I wish to connect these expressions to particular ideologies relevant to environmental education and to give a few examples of the language arts that apply to each. This does not mean that I am presenting a model that frames the whole of the field of the environment. Neither am I offering a chronologically based commentary which might erroneously suggest contemporary society is best at revealing how to safeguard the environment. And I am not suggesting that the metaphors below are entirely distinct from each other. I suggest that these few metaphors provide an initial mechanism or starting point for understanding the choices that we have in the characterisation of the environment. Then, in a later section of this chapter, and using the ideas of the contributors to this book, I develop a particular way of characterising educating about, through and for the environment and language arts.

Paradise

Perhaps the metaphor of paradise is the most fundamental of all characterisations of the environment. The perfection of idealised sites from a range of cultures (such as Eden, from a Christian perspective) is corrupted by knowledge with self-awareness and rationality leading to the guilt, shame and nastiness of agentic life. The attraction to the beauty of nature, the sense of wonder that is good for us and its contrast with the limited capacity of what we alone are capable of is shown by very many writers of which perhaps the Romantic poets are the most obvious example. The sense that rationality is not as powerful as a spiritually informed awe may be seen, for example, in Wordsworth's *Intimations of Immortality,* in Thoreau and Coleridge and while some attention may be paid to the more challenging aspects of the relationship between naivete and understanding (here, the motto to Blake's innocence and experience is instructive), the overpowering impact of this approach to nature is a way of seeing the deity in all things.

The fearful symmetry

My second metaphor is not necessarily the polar opposite of paradise, but it is a distinctive aspect of many approaches to and understandings of nature in which we are not only in awe of beauty but experience terror. Blake's tiger is not only beautiful. The poem *Design* by Robert Frost points to the unpleasantness of aspects of nature and questions the link with the divine. Dorothea Mackellar's love for Australia is dependent upon its violent climate. No one could see Ted Hughes' work as offering a cosy view of nature. The wildness and savagery of the natural world may be seen in *Beowulf* and many other ancient classics. Nature, as Tennyson would have it, is red in tooth and claw.

The inexhaustible gift

There are two aspects to this metaphor: a never-ending supply of resources and the availability of dumping ground that will take the waste from an extractive manufacturing process. There are very many novelists, playwrights and poets who explore this aspect of nature. Most obviously are those associated with imperialism and colonialism. Kipling (despite attempts to re-interpret the nature of his work) seems aligned with this stance as well as R.M. Ballantyne and many others. The presentation of the destruction of wildlife may be seen in such work as J. M. William's *Butcher's Crossing*. The consequences of the never-ending commitment to growth becomes associated with the literature of catastrophe (in benign forms such as Clive KIng's *Stig of the Dump*, and in more frightening ways in for example Cormac MacCarthy's *The Road*).

Real utopias (Torres, 2023)

Like many other contrbutors to this book, I am very much drawn to the work of Robert Macfarlane (2019) who uses the phrase "radical re-storying" to characterise a new approach to the environment and literature. The key ideas here are about seeing the relationship between people and the environment as integrated, interdependent and dynamic. Philosophically, the work of Weil (2023) is relevant, seeing the necessary connection between light and growth and as such being able to represent the roots of a tree as being in the sky. Very many authors (for example: Powers, 2018) may be cited to illustrate the sense of the environment being not distinct from ourselves.

Of course, it is not for me to declare the precise characterisation and perspective held by the authors of this book and the metaphors themselves are not mutually exclusive, but there may be some sense of attachment to the last of the metaphors I have outlined above. There is great value in the sense of the language arts as a means to engage and not just to use, to embrace and not to separate us from the natural world. I am struck by Fiona Mozley's image from her novel *Elmet* in which "The soil was alive with ruptured stories that cascaded and rotted, then found form once more and pushed up through the undergrowth and back into our lives" (British Council, 2023).

Connecting characterisations of the environment with political ideologies

My point in highlighting these metaphors is to argue that how we view the environment connects with what matters to us and shapes what we write in the language arts and thus how we educate. Five themes can be highlighted from the metaphors to illustrate the political meaning of approaches to nature and language. The first two of these themes are, to my mind, straightforwardly negative, while the final three themes have more positive potential for a socially just form of sustainability but also bring significant challenges to our sense of how we should think and what we should do.

Firstly, while there is much to celebrate in the wonder of the natural world, there is a possible connection between some of the metaphors outlined here – especially paradise – and an attachment to the idea of purity. This may not be a benevolent position. Bramwell (1994) has discussed the connections between desire for purity and extreme right-wing, fascist political stances. I am not (of course) accusing those who are contemporary climate activists of unacceptable political positions, but I am declaring the need for careful thought about the various ways in which the roots of environmentalism may be explained.

My second theme about the meaning of environmentalism is associated with the impact of 19th century industrialisation and scientific developments. These developments shocked writers away from a spiritual awe about nature and towards a Darwinian desire to conquer the landscape (Schama, 2004). This connects with particular political ideologies in that the new-found sense of people's power over nature is not separate from the rush to create colonial empires. Even the emphasis on human rights from the beginning of the modern world (since approximately 1789) is interpreted by some libertarians as allowing for the possibility for individuals to do what they want – including, for example, slaughtering animals to extinction. This libertarian approach to society is, whether we like it or not, part of debates about a rights-oriented civilisation. If we refuse to think openly about the roots of ideas and actions, then just as in Adam McKay's movie *Don't Look Up*, environmental collapse is brought closer.

If we can (I hope) immediately reject any ideas that connect with fascism, colonialism and imperialism, what are the consequences of what I would consider to be much more appropriate, much more valuable ideas about the environment? Even in a more ameliorative context, there are challenges. My third theme – or, difficult challenge – about the way to view the environment and how that might connect with political ideologies, arises from an extension to the debates about rights in the work of Macfarlane (2019) and others. There is now important mainstream philosophical work that debates, in a relationship between people and the environment that is interdependent, whether nature itself has rights (NECE, 2023b). If we are to accord due regard to an environment that is interdependent with people then that idea has significant consequences. If, as an example, we were to imagine and enact rights for the environment (as has been done in certain places, such as Toledo, Canada, in 2018 regarding a bill of rights for Lake Erie) then how might society be characterised and ordered? Macfarlane (2019) asks: If a tree has rights, where is the beginning and end of a forest? If a river has rights then are there corresponding responsibilities which could lead to action being taken during floods? and, if we were to restrict rights to those beings which we regard as sentient (principally, animals) would it be appropriate if this pertained only to those that are, at least in some ways, similar to, or useful to humans?

These challenges about rights connect with my fourth theme, which is the emphasis which we feel may be legitimately placed on the usefulness of the environment. Utilitarianism brings significant challenges (see discussion of the work of Singer by Gubler, 2023). If we emphasise the greatest happiness of the greatest number, then would it be acceptable, for example, for cruel treatment to be meted out to animals in order to feed people or would a useful regard of animals mean that humans (for example, in the evacuation of a war zone) could not expect special treatment? Is it acceptable to evacuate a dog from Afghanistan and to leave people behind (Graham-Harrison & Gentleman, 2022)? Generally, the rise of the effective altruism movement (MacAskill, 2016) is not without significant challenge unless one is persuaded of the value of utilitarianism. Finally, my fifth challenge is to be found by turning to Aristotelian arguments for virtue (or, simply, doing the right thing) which is explored in this book at various points including in Chapter 6 with valuable reflections on the need for philosophical dialogue. That, surely, is the only thing to do. But that is the only thing to do *if* we could find what that actually means and if we know how to adopt a process that allowed us to get to that point. We need much more than virtue signalling.

All the authors of this book are engaged with the characterisation of nature, are alert to the political ramifications of such work and are keenly exploring the ways to make their ideas meaningful through the language arts. The different views that we hold of the natural world and the struggle over the connections that necessarily exist with how we achieve what we really value is not straightforward. The language arts are perfectly positioned to allow us to explore these matters and by so doing approach a socially just characterisation and practice about and for the environment.

Learning and teaching about, through and for the environment

What then do these discussions about the environment and language arts mean for education? There are many very different perspectives on a curriculum (including whether it is intended, enacted and/or experienced) and, of course, we need to be alert to what lies behind its formation. The authors of this book make their own cases for how language arts may play a role in environmental education. For serious attention to be paid to the environment, all three of the drivers mentioned by Ross (2000) are relevant. Firstly, "content" may be included to allow for a critical examination of the best that has been said, thought and done. This liberal perspective is designed to promote rationality within a particular cultural context and is associated with philosophers such as Hirst (1965/2010). Secondly, some suggest that "objectives" must be recognised. This approach is associated with, for example, Dewey (1916/1966) and may be clearly seen in this book in Echterling's (Chapter 2) argument for connecting environmental degradation and social inequities. Policymakers often fail to make aims clear (see for example the fact that since September 2022 and up to at least the time of writing (November 2023) England became the only one of the four UK nations not to have an aims-based school curriculum (White, 2022). Thirdly, the need for environmental education emerges from those who argue for "process" (associated fundamentally with Rousseau's *Émile* (1762/1991), and also the progressive child development movement of the 1960s and 1970s, which in some forms may still be seen). Sandel (2009) links these three educational drivers to political perspectives, allowing us to reflect on what is at stake in establishing a curriculum. It is rarely (if ever) the case that one perspective is completely dominant but my point here is that as we frame our learning and teaching about, for and through the environment, we are declaring (even if only implicitly) the nature and purpose of fundamentally important matters.

An essential part of establishing the educational process is to move from the drivers of curriculum to how the environment and the language arts may be framed. There is a debate about whether there is a need for discrete traditional subjects with students given the intellectual and other tools to make the connection with the environment; or, whether the environment should be deliberately highlighted by infusing key ideas and issues through those subject carriers; or, whether there is the need for a radical reform of what we currently offer in schools, perhaps through the development of a cross-disciplinary approach as described for example in Chapter 13. We can also see such themes being explored in, for example, Braund's arguments in Chapter 3 for the humanisation of science education as well as in Küchler's approach to additional language learning in the context of environmental education (Chapter 4). To make a success of these intentions we are being usefully prompted to consider how the interplay between the environment and the language arts is developed. Bernstein (1971) argued that the degree of specificity about classification (in other words, the distinctions between areas/subjects) and form (in other words, the fixity of the relationships between those involved in a teaching/learning process) would need to be constructed so as to allow for creative dynamism. It is possible that the titling of subjects, themes or whatever we choose to call them and the ways those things relate to the arguments about disciplinary, interdisciplinary, cross-disciplinary and multi-disciplinary work need not be driven to some sort of one size fits all curriculum. There is, of course, to refer again to the chapters of this book, useful overlap within the language arts between what could be described, loosely, as the tools needed to engage (such as reading, writing and voice in a first and/or additional language) and the fields within which engagement occurs such as drama (Chapter 10), the visual image (Chapter 12) and literature (Chapters 1, 2, 7 and 8) as well as those areas through which the language arts may be infused (such as science, children's geographies) and the institutionally contextual connections (such as teacher education and in international locations). All areas, in order to be valid in a positive learning environment, need to be both sufficiently distinct to allow for coherent meaning and sufficiently plastic to allow for creativity. All "subjects" are to some extent meaningful entities that carry status and they also exist as hybrids. The thrust of the argument here is that we need to know what we are planning to learn and teach and why, and we need to be alert to arguments about knowledge and understanding. It seems reasonable that the environment is sufficiently important to be seen as a body and form of knowledge and not just a topic.

Once these drivers and framing devices have been made clear, we need to go further and elaborate the sort of outline that would be appropriate for language arts and the environment. There is a need to go beyond the stale and inert lists of content that students are at times required to remember and regurgitate in formal examinations. Pike and Selby (1988) in a classic work on global education emphasised health of planet awareness and stressed the need for systems consciousness, perspective consciousness, involvement consciousness and preparedness and process consciousness. Marshall (2005) and others refer to the need for a "global gaze", suggesting something about the arts and the environment which has a different perspective from a simple – and perhaps simplistic – bureaucratic codification. Beyond general considerations to do with the development of the curriculum, there is a need to identify the substantive concepts (what is this "subject" or "area" about?) and procedural concepts (how must this "subject" or "area" be studied in order to achieve learning?) (see Hudson & Shelton, 2021). At the heart of an approach to the environment and the language arts should be a conceptual framework. Concepts as the bedrock of education allow us to think and to act, going beyond lists of content, context, skills and virtues. Reliance on key concepts provides for engaged constructivism in which there is clarity and, importantly, an appropriate purposefulness about what we are learning and how we are learning it. If the environment is to be viewed as a living force, in which it – and we – are dynamically and interdependently engaged then we have to develop a suitable conceptual bedrock. There is a need for very careful reflection and the development of various iterations of what that bedrock is. It would be remiss not to begin an initial sketch of what that interlocking framework could possibly look like. The substantive concepts might consist of place (where we live), socially just ideas and practices (how we live) and identity (who we are). The procedural concepts could be characterised by understanding (comprehending and interpreting the environment); critiquing (being able to develop one's voice and advocacy in relation to the environment); and engaging about, in and through the environment (acting for educational purposes, reflecting on what, why and how those actions were conceived, developed and received). It is vital to see the connections between these things as well as their distinctiveness. So, for example, unless we are to unintentionally require rote learning, we must characterise understanding, necessarily, in part as critique. The procedural concepts are similarly interlinked. They are, broadly, sketched as thinking, speaking and acting and, for example, voice is a part of action, to think is to begin the process of advocacy and so on. The language arts are particularly suited to such a conceptual framework. The forms of drama, (visual) imagery, literature, creative writing and the practices of reading, writing and performance explored in the various chapters of this book, provide an essential dynamism that breathes life into the substantive and procedural. These forms and practices are, simultaneously, illustrative of each of the substantive and the procedural and provide a bridge between them. And, again, the links between these things are vital. To take one example: drama (Chapter 10) is both a form and a practice. Given the osmotic interplay of what is outlined here, I have deliberately chosen not to display the framework in the form of a static figure or diagram.

If we are the environment and we need to learn about it, through it and for it, then we need to develop a framework for learning that emerges from that characterisation. And in doing so, we need to provide something that is sufficiently generic and transferable as the language arts are for all and are in all that we do, and, importantly, something that is appropriately specific, that recognizes the particular ideas and practices of what can and should be done through the contribution that the language arts make to environmental education.

What is to be done?

In this final main section of the conclusion, I will, briefly, highlight aspects of education that I feel the chapters of this book are prompting us to explore further. I will mention briefly three key areas: pedagogy; assessment; and the support that is needed for the language arts to contribute to environmental education.

Pedagogy and political action

One of the many strengths of this book is that the authors of chapters have indicated approaches to teaching and learning in thoughtful, practical and exciting ways. For example, the work outlined in Chapter 13 to develop science literacy, and on the concept of "taking action" within the context of EFL teacher education (Chapters 4 and 11), gives the reader a clear outline of what can be done. The authors bring to life the discussion about what sort of action may a student take and for what purpose. This links with the many good sources of guidance for teachers produced by Oxfam, the Association for Citizenship Teaching, the Historical Association and many others. The key question for teachers is what is being learned when students are engaged in political action, rather than what is being achieved in the form of a political objective (although, of course, there will almost always be a strong overlap between these things if we are to avoid involving young people in activities that may be politically harmful to them).

Beyond the formal and explicitly declared curriculum, there are important considerations about learning within any educational relationship. The so-called "hidden curriculum", the relationships between people, the issues of power and authority are centrally relevant to the environment and the language arts. Who speaks and to what effect must be recognised as being of central importance. There are ways (see Chapters 1, 7 and 10 for example) of avoiding the "arrogance of English speakers, the way they contrived, by virtue of their tongue, to bring destruction with them, wherever they went in the world" (Bernstein, 2023, p. 64). At times, the achievement of a good relationship will be dependent on structures within the school. It is valuable, for example, to have a mechanism such as a school council, to allow students real power, exercised in a manner that is educational. The ways in which the school interacts with the local, national and global communities of which it is an essential part are all important and must include work that takes place across and beyond boundaries (see for example Chapter 13).

Assessment

There are significant dangers in ideas about assessment that rely on a form of meritocracy that ignores valuable learning and denies the common good (Sandel, 2020), devalues the role of the teacher in assessing coursework and impacts negatively on student health (Lee & Larson, 2000). But assessment is both essential and unavoidable. Against those who would argue that assessment per se is inappropriate for investigations into and practices for the environment, I am suggesting that a good teacher obviously (even if at times only implicitly) is interested in what sort of learning is taking place. Assessment not only gives an area status, it is an essential part of the dialectical process between and within teachers and learners. The students are as interested in self and other forms of assessment as the teacher, needing to know what sort of progress they are making and making judgments about the extent to which the teacher is helping them to their desired goal. Of course, assessment in the form of closed tests designed to give summative results about whether a student can memorise information has extremely limited value.

General support

The environment will remain a rhetorical device unless there are specific initiatives to support it. There is a need for carefully framed policy documents to be created. Official government initiatives may be extremely unhelpful, but there may be positive potential from legislative statements that invite creative and thoughtful action. Also essential - and very often emerging from the prompt supplied by governments (but also from grassroots activism) - will be those NGOs and professional associations that can do so much for raising the status and providing practical guidance about what is and could be done. It will be of enormous benefit if these things are prompted by and enhanced by research. There may be evaluations which help teachers and students to focus their work. And, at every level of the educational process, there is the need for the education of teachers (initial and continuous professional development – see Chapters 9 and 11, for example).

Conclusion

At a time of unprecedented existential crisis, the authors of this book have made "a case for the role which language arts might play in environmental education". They have explored and clarified the nature of environmental education, language arts and the relationship between them. The achievement of that clarity on the basis of careful thinking and the review of practical initiatives provides not only a very valuable way forward for academics, professionals and all those who care about and who are involved in education in its broadest sense, but also is an invitation for more work to be done. And there is a considerable amount of work that is needed now and in the future. Many key stakeholders, remarkably, remain to be convinced of the urgent need for necessary forms of education. There must be an urgency in the development of educational initiatives and in the establishment of creative support structures for that work. The aim must be for the environment to flourish. If action is not taken, the arts will document its demise.

References

Bernstein, B. (1971). On the classification and framing of educational knowledge. In M. F. D. Young (Ed.), *Knowledge and control: New directions for the sociology of educational knowledge* (pp. 47–69). London: Collier-Macmillan.

Bernstein, S. (2023). *Study for obedience.* London: Granta.

Bramwell, A. (1994). *The fading of the greens. The decline of environmental politics in the west.* New Haven, CT and London: Yale University Press.

British Council. (2023). Literature. Retrieved November 23, 2023, from https://literature.britishcouncil.org/writer/fiona mozley#:~:text=In%20the%20words%20of%20Danny,are%20both%20conspicuous%20and%20embedded

Buell, L., Heise, U. K., & Thornber, K. (2011). Literature and environment. *Annual Review of Environment and Resources, 36,* 417–440. http://dx.doi.org/10.1146/annurev-environ-111109-144855

Carnegie Europe. (2021). A global citizens' assembly on the global and ecological crisis. Retrieved November 23, 2023, from https://carnegieeurope.eu/2023/02/14/global-citizens-assembly-on-climate-and-ecological-crisis-pub-88985

Department for Education. (2022). *Political impartiality in schools.* Retrieved November 23, 2023, from https://www.gov.uk/government/publications/political-impartiality-in-schools/political-impartiality-in-schools#:~:text=It%20is%20important%20to%20remember,to%20teach%20about%20controversial%20subjects

Dewey, J. (1916/66). *Democracy and education.* London: Free Press.

Fredericksen, E. (2019). The environmental poetry of Augustan Rome. (Doctoral dissertation, Princeton University – Princeton, NJ). Retrieved November 18, 2023, from http://arks.princeton.edu/ark:/88435/dsp018336h4833

Gibbs, Jr., R. W. (2011). Evaluating conceptual metaphor theory. *Discourse Processes, 48*(8), 529–562. https://psycnet.apa.org/doi/10.1080/0163853X.2011.606103

Graham-Harrison, E., & Gentleman, A. (2022, January 26). It's a joke. Left-behind Afghans despair at dog rescue revelations. *The Guardian.* Retrieved November 18, 2023, from https://www.theguardian.com/world/2022/jan/26/its-a-joke-left-behind-afghans-despair-at-dog-rescue-revelations

Gubler, S. (2023, November 10). Ethics without humanism: New editions of the philosopher Peter Singer's classic works. *Times Literary Supplement,* pp. 10–11.

Hirst, P. H. (1965/2010). Liberal education and the nature of knowledge. In R. D. Archambault (Ed.), *Philosophical analysis and education volume 1* (pp. 76–95). Abingdon and New York, NY: Routledge.

Huckle, J. (2012). Sustainable development. In J. Arthur & A. Peterson (Eds.), *The Routledge companion to education* (pp. 362–371). Abingdon and New York, NY: Routledge.

Hudson, B., & Shelton, C. (2021). The curriculum: Developing powerful knowledge and creative know-how. In B. Hudson, M. Leask, & S. Younie (Eds.), *Education system design. Foundations, policy options and consequences* (pp. 141–153). Abingdon and New York, NY: Routledge.

Kemp, S. (2023, October 11). Education for sustainable development: Engaging with staff, quality assurance, curricula and the student experience. *Sustainability Education Conversations* (by zoom), University of York.

Lee, M., & Larson, R. (2000). The Korean 'examination hell': Long hours of studying, distress, and depression. *Journal of Youth and Adolescence, 29,* 249–271. https://psycnet.apa.org/doi/10.1023/A:1005160717081

Lewis, M. (2018). *The fifth risk.* London: Allen Lane.

MacAskill, W. (2016). *Doing good better.* New York, NY: Avery Publishing Group.

Macfarlane, R. (2019, November 2). Should this tree have the same rights as you? *The Guardian.* Retrieved November 18, 2023, from https://www.theguardian.com/books/2019/nov/02/trees-have-rights-too-robert-macfarlane-on-the-new-laws-of-nature

MacLeod, D. (2007, October 10). Climate change film to stay in the classroom. *The Guardian.* Retrieved November 18, 2023, from https://www.theguardian.com/education/2007/oct/10/schools.uk

Marshall, H. (2005). Developing the global gaze in citizenship education: Exploring the perspectives of global NGO workers in England. *International Journal of Citizenship and Teacher Education, 1*(2), 76–92.

McAuley, T. (2023). *Seeds in the heart and leaves of words. Traditional Japanese poetry beyond the Haiku.* Presentation 7 December at Japan House, London, Japan Foundation.

NECE (Networking European Civic Education). (2023a). A portrait of the artist as a civic educator? Retrieved November 23, 2023, from https://nece.eu/a-portrait-of-the-artist-as-a-civiceducator/?utm_source=brevo&utm_campaign=NECE%20NEWS%203rd%20issue%202023&utm_medium=email

NECE (Networking European Civic Education). (2023b). Re-imagining civic education by cultivating more-than-human-care. Retrieved December 18, 2023, from https://nece.eu/nece-campus/

Pike, G., & Selby, D. (1988). *Global teacher, global learner*. London: Hodder and Stoughton.

Powers, R. (2018). *The overstory*. New York, NY: W. W. Norton.

Ross, A. (2000). *Curriculum: Construction and critique*. London: Routledge.

Rousseau, J. J. (1762/1991). *Émile or on education*. London: Penguin.

Sandel, M. (2009). *Justice. What's the right thing to do?* London: Penguin.

Sandel, M. (2020). *The tyranny of merit. What's become of the common good?* London: Allen Lane.

Schama, S. (2004). *Landscape and memory*. London: Harper.

Torres, A. (2023). Global citizenship education and sustainability as real Utopias. In D. Bourn & M. Tarozzi (Eds.), *Pedagogy of hope for global social justice: Sustainable futures for people and the planet* (pp. 13–29). London: Bloomsbury.

United Nations. (2023). Climate change. Retrieved November 23, 2023, from https://www.un.org/en/climatechange

Wals, A. (2009). *Review of context and structures for education for sustainable development*. Paris: Unesco.

Weil, S. (2023). *The need for roots. Prelude to a declaration of obligations towards the human being*. London: Penguin Classics.

White, J. (2022). The aims of the curriculum should be the fount from which everything else in school life should flow. IOE Blog. Retrieved November 23, 2023, from https://blogs.ucl.ac.uk/ioe/2022/04/22/the-aims-of-the-curriculum-should-be-the-fount-from-which-everything-else-in-school-life-should-flow/

Index

Note: Page numbers in *italics* indicate figures, bold indicate tables in the text, and references following "n" refer notes.

Made in the USA
Las Vegas, NV
19 September 2026

97e5de55-9dab-445f-a136-2abb8978cc96R01